SIMULATION

Sociology of the Sciences Yearbook

VOLUME XXV

The titles published in this series are listed at the end of this volume.

SIMULATION

Pragmatic Construction of Reality

Edited by

JOHANNES LENHARD
University of Bielefeld, Germany

GÜNTER KÜPPERS
University of Bielefeld, Germany

and

TERRY SHINN
GEMAS Paris, France

A C.I.P. Catalogue record for this book is available from the Library of Congress.

ISBN-10 1-4020-5374-6 (HB)
ISBN-13 978-1-4020-5374-0 (HB)
ISBN-10 1-4020-5375-4 (e-book)
ISBN-13 978-1-4020-5375-7 (e-book)

Published by Springer,
P.O. Box 17, 3300 AA Dordrecht, The Netherlands.

www.springer.com

Printed on acid-free paper

TABLE OF CONTENTS

INTRODUCTION

Chapter 1: COMPUTER SIMULATION: PRACTICE, EPISTEMOLOGY, AND SOCIAL DYNAMICS 3
Günter Küppers, Johannes Lenhard, and Terry Shinn

IMITATING MODELS

Chapter 2: THE SHAPE OF MOLECULES TO COME 25
Ann Johnson

Chapter 3: FROM REPRESENTATION TO PRODUCTION: PARSERS AND PARSING IN LANGUAGE TECHNOLOGY 41
Tarja Knuuttila

Chapter 4: FOUNDATIONS FOR THE SIMULATION OF ECOSYSTEMS 57
Michael Hauhs and Holger Lange

Chapter 5: MODELS, MODELS EVERYWHERE 79
Don Ihde

LAYERS OF INTEGRATION

Chapter 6: FROM HIERARCHICAL TO NETWORK-LIKE INTEGRATION: A REVOLUTION OF MODELING STYLE IN COMPUTER-SIMULATION ... 89
Günter Küppers and Johannes Lenhard

Chapter 7: THE DIFFERENCE BETWEEN ANSWERING A 'WHY' QUESTION AND ANSWERING A 'HOW MUCH' QUESTION 107
Marcel Boumans

Chapter 8: STRUGGLE BETWEEN SPECIFICITY AND GENERALITY: HOW DO INFECTIOUS DISEASE MODELS BECOME A SIMULATION PLATFORM? 125
Erika Mattila

Chapter 9: HANDSHAKING YOUR WAY TO THE TOP: SIMULATION AT THE NANOSCALE 139
Eric Winsberg

SOCIAL PRACTICE

Chapter 10: LOCATING THE DRY LAB ON THE LAB MAP 155
Martina Merz

Chapter 11: SIMULATION UNCERTAINTY AND THE CHALLENGE OF POSTNORMAL SCIENCE 173
Arthur C. Petersen

Chapter 12: WHEN IS SIMULATION A RESEARCH TECHNOLOGY? PRACTICES, MARKETS, AND LINGUA FRANCA 187
Terry Shinn

LIST OF AUTHORS 205

BIOGRAPHICAL NOTES 206

NAME INDEX 209

INTRODUCTION

CHAPTER 1

GÜNTER KÜPPERS,* JOHANNES LENHARD,* AND TERRY SHINN**

COMPUTER SIMULATION: PRACTICE, EPISTEMOLOGY, AND SOCIAL DYNAMICS

What does the word 'simulation' refer to? What is done during a simulation, and what are the technical, intellectual, and epistemological issues raised by it? Who are the practitioners of simulation? What sorts of problems are addressed? What is the scope and composition of the market? Finally, if anything, what does simulation have to do with transformations in science, in technology, and, if postmodern thinkers are to be believed, in the very structure and substance of contemporary society? This book attempts to address some of these questions, and in doing so, it often raises additional ones.

The word 'simulation' comes from the Latin *simulare*. For almost three centuries, the principal lexical meaning of simulation in the English, French, and German languages referred to 'imitation' or, alternatively, to 'deception.' In everyday parlance, someone simulates when he imitates a certain behavioral pattern, for instance, the actor in a drama, but also a malingerer who imitates the symptoms of a disease, in an authentic, albeit deceitful, way. A case from literature is Felix Krull, from Thomas Mann's novel *Confessions of Felix Krull, Confidence Man* (1954). Krull studies medical literature to learn about the symptoms of a particular nervous disease, and subsequently simulates the disease to deceive military doctors and obtain a medical exemption from the army. A slightly different meaning of simulation is equated with illusion: In late Renaissance and Baroque painting, the imitation of tableau became fashionable. One famous example is a painting by Cornelius Gijsbrechts (about 1670) entitled *Back of Painting* (see Figure 1), which seems to depict what the title says. The spectator's impression is of a real painting hanging on the wall, but showing the back of the canvas. This example of illusionistic painting in fine art may also count as an instance of simulation.

The meaning of the term simulation changed after World War II, as the definition given by the *Oxford English Dictionary* (fourth edition 1989) reflects: "The technique of imitating the behavior of some situation or process [...] by means of a suitably analogous situation or apparatus, especially for the purpose of study, or the training of personnel." In contemporary life, however, simulation has generally come to

J. Lenhard, G. Küppers, and T. Shinn (eds.), Simulation: Pragmatic Construction of Reality, 3–22.

Figure 1. Cornelius Gijsbrechts: Trompe l'oeil. The reverse of a framed painting. (By courtesy of the Statens Museum for Kunst, Copenhagen. Photographer: SMK Foto.)

be equated with science and technology and is viewed as synonymous with computation and the digital computer.

In recent decades, simulation has increasingly become established as a new means of knowledge production and especially representation of complex dynamics in science and technology as well as a tool for the development of new and better technical artifacts in a rapidly expanding range of fields. Undoubtedly, one essential reason for this development is the amount of computing power that has become available over the last twenty-five years, and it is perhaps not inappropriate to think of simulation as 'computer simulation,' so strongly connected is simulation to the computer and computer science. The diversity of the sites of usage, applications, and practitioners connected with computer simulation today have turned it into a pervasive and often prominent social, organizational, and cognitive sphere that either directly or indirectly, unwittingly or consciously, impacts on the lives of most people.

Computer simulations are applied in science, technology, engineering, different areas of technical and professional training, economics, leisure, and art. To illustrate the broad field of applications, we cite three examples: In science, the dynamics of galaxies, encompassing billions of stars, cannot be grasped theoretically or experimentally. The fundamental theories are known and unquestioned, but the resulting mathematical equations cannot be treated by the traditional analytical methods. Computer simulation is currently viewed as the sole acceptable path for exploring a complex universe. In technology and engineering, the situation is similar. The investigation of how colliding cars behave and how passengers become injured can be conducted in experimental crash tests. Yet many automotive companies prefer virtual collision tests conducted during the R&D phase rather than awaiting experimentation using advanced prototype vehicles. Finally, climate change has become a major issue in science, in policy, and in the media. What will be the consequences of global warming? Computer simulations are the main instrument for obtaining predictions here as well.

Traditional scientific knowledge has generally taken the form of either theory or experimental data. However, where theory and experiment stumble, simulations may offer a third way. The central question is: What are the characteristics of this mode, and how reliable is simulation-based knowledge? If computer simulations provide a new way beyond theory and experiment, that is, if they are not merely numerical solutions of theoretical problems, new practices of validation and assessment also become necessary. Alternatively, the roles of simulation within science may prove more restricted, and its epistemological effects more limited.

It is important to ask: Does simulation constitute a newly emergent scientific discipline? There exist over a score of scholarly journals in the Science Citation Index database specifically connected to simulation; yet does this necessarily signify that simulation should be regarded as a scientific or technical discipline? Is this number of reviews as elevated as one might anticipate for a 'revolutionary' full-fledged research domain? Indeed, it proves extremely difficult to identify the social and organizational locus of computer simulation. There are no university departments in the field, no diplomas, no established intellectual corpus, or certified body of skill. But does it necessarily follow that in terms of social and organizational significance, simulation represents nothing more than a merely loosely coupled, fragmented body? It may be queried whether simulation is not instead a historically important, perhaps even historically unusual, research instrument. One thing is certain, simulation is a relatively new entity, whose usages are in flux and whose 'good practices' have not yet even been determined in full.

Computer simulation is a domain of growing interest to sociologists, historians, and philosophers of science. Sociologists query the organizational and material conditions that surrounded simulation's foundations, question the dynamics and structure of the movement, interrogate the internal form of the occupation/profession, and focus on its relations with other bodies as well as the size and scope of its market. They are concerned with the shape of the computer simulation field, the expression of its diverse forms of symbolic capital, the forms and rules of competition, what counts as legitimacy, and finally, they are concerned with the relations between the field of simulation and other science, technology, and fields beyond (Bourdieu 1975, 2001). For their part, historians of science demand to know the backdrop of simulation activities; who practiced it; where, why, and how. To what extent does computer simulation constitute an extension of earlier practice and forms of knowledge, and to what extent does it comprise something unprecedented? Finally, due to the complex and ambiguous linkage between simulation, models, and representation, philosophers of science too are increasingly drawn to this often elusive domain. They are interested in the epistemology and methodology of simulation and also in the complex relations extant between theory, models, simulation models, computation, and the material laborant to which they all refer. In order to frame a clearer understanding of the aforementioned problems, this book assembles contributions from the intersection of all three domains.

GENESIS AND BACKDROP

Prior to the appearance of simulation in science, itself now linked to digital computers (and for that matter even to any form of computer), a kind of simulation was already applied in technology. In the late nineteenth century, nautical design was sometimes assisted by data and ideas obtained by studying the behavior of miniature ship hulls carefully displaced through a variety of hydraulic conditions. Development of such early simulation was stimulated by the passage from sail to steam and from wood to steel. Traditional knowledge about wooden hulled sailing boats had been outdated by iron as new materials for the construction of bigger and faster ships developed. Experiences with the new steamboats were rare.

This early real-world simulation may be associated with a form of early technical modeling that differed from previous practices based on the extension and modification of noncodified craft data and on lessons drawn from observing unfortunate design errors. France had a different nautical tradition based on applying mathematics and deductive principles to ship building. However, this often remained disconnected from observational inputs. At that time, the theories to describe the relation between the resistance of a body in water flow and its velocity were available to physics. However, the resulting equations had no general solution because of nonlinearities. Hence, when investigating the influence of different hull shapes, one had been limited to trial and error – a costly affair with full-size ships. Later on, the wind tunnel was employed as a simulation instrument to investigate the dynamic properties of objects in air flow in a very similar way. It may reasonably be hypothesized that the form of simulation practiced during this era may have acted as a sort of bridging mechanism that drew diverse and divergent design practices more closely together.

The twentieth century witnessed a huge growth in the frequency of this kind of 'real-world' simulation that takes place in reality and not in the symbolic realm of a digital computer. Already in the interwar era, simulation had been proposed and developed for the solution of technology-related problems. In 1929, German engineers took out patents for a device designed for training pilots in airplanes, dirigibles, and submersibles. The apparatus involved elementary indicators of vehicle altitude, an altitude control system, and an interactive system between the two mechanisms based on electromechanical devices. Response flight simulators permit the training of pilots who have to react correctly in risky situations – without risking a 'real' crash. Throughout World War II, simulated flight and gunnery training became common. In the later stages of the war, physicists and engineers sometimes managed to harness analog computers to simulation, with astounding consequences. The introduction of the computer permitted critical advances on three fronts: (1) Simulated experience became more 'realistic' due to finer-grained responses and shorter response time. (2) More situations and variables could be introduced. (3) The new capacity to inject information into simulation based on the real-time solution and representation of complex mathematical equations not only refined simulated learning but also transformed simulation into a research tool. Very soon, simulation moved beyond training and became a central instrument in technical design, particularly for aircraft and rocket development. One perceives here the genesis of a virtual simulation cycle in which the 'reality-constrained' feaures in technology simulation fuel and advance the 'symbol-bounded' features in science simulation; and the symbol-bounded methodologies, representations, and proofs of simulation in science nurture the realities embodied in technological simulations.

The advent of the digital computer triggered a radical transformation that changed simulation from a refined technology for imitation into a full-scale polyvalent research instrument. Nonetheless, at first glance, the shift within simulation might appear to be rather trivial and mainly technical, constituting an important advance, but not a decisive one. In flight training, for example, analog devices were replaced by digital computing devices. Yet, despite this technical substitution, for all ostensible purposes, the flight simulator remains a flight simulator. However, this seeming invariance obscures a fundamental discontinuity. The transition from analog devices to digital simulation models, which, for example, describe the dynamic behavior of a plane's wings, transformed the very essence of even the flight simulator by enabling it to generate physically possible, even likely, aircraft performance, which to date had not yet been observed. In effect, the flight simulator commanded by a digital computer is capable of extending a vehicle's latent material conditions and the scope of pilot experience beyond observed routines. The meaning of simulation is thereby deeply transformed. This book is devoted to digital computer simulation. It will focus on the new aspects introduced through computer simulations, distinguishing them from older usages.

The student of the practices, epistemology, and social dynamics of computer simulation wants to know how and why this important transformation came about. Was it connected with the introduction of new problems, or even a new species of problem on the research agenda that could not be examined other than by simulation? Did the acceptance and spread of simulation in science signify the introduction of some new, commonly accepted form of proof of the reliability of simulation outputs? Does simulation represent a general switch, whereby a younger generation sets itself apart from older generations through the adoption of a formerly low-status and little used technique? And, beyond all this, can the prevalence of simulation in science today be likened to a 'paradigm shift': Does it necessarily entail the emergence of a new way of knowledge production incommensurable with the common ones (that is, theory and experiment)? Or more conservatively, is simulation instead mainly a tremendously powerful generic instrument, constituting an enabling device? These questions themselves reveal that simulations mark a multifaceted change, as indicated by the following four interacting factors:

1. The pace of evolution in the speed and capacity of calculation in computer technology (Humphreys 2004) obtained through the technological development of hardware and software makes increasingly complex problems accessible. The steep increase in speed and quantity is an important determinant of the possibilities and limiting conditions of simulation as an instrument. Developments in high energy physics (Merz, this volume) and in economics (Boumans, this volume) document how the availability of the computer as a technological instrument has opened up new fields of application that have, in turn, permanently driven the

scientific characteristics of simulation. On a slightly different register, the total reliance of nanotechnology research on the computer demonstrates, for instance, that computer simulations go far beyond simply generating a mutual adaptation between science and the computer: The computer has changed the very nature and form of the questions being asked in this field and has transformed the models being constructed (Johnson and Winsberg, this volume). The technology of the computer is by no means fixed, and with increasing computing power, things change decisively.

2. This development is connected closely to the capacity to generate visualizations, to process images, or more generally, to handle ever more sophisticated man-machine interfaces. Computer images render visible the fine-grained details of atoms (in nanoscience, see Johnson and Winsberg, this volume) as well as the global dynamics of the climate (both in Technicolor). Such graphics underline the character of simulation as an 'observational instrument,' but one in which the concept 'observation' assumes an entirely novel meaning. They can enable access to complex patterns of behavior undetected by classical instruments such as telescopes or microscopes. Whereas telescopes and microscopes render phenomena visible by affecting the *scale* of 'tangible' entities through optical processes of resolution, simulation renders 'visible' the affects of parameters and forces such as time, dynamic interactions, and so forth that are not dealt with by optics-related transformations. Thus, simulation, by constructing images, may translate absolutely nonvisual events into a visual media! Often there is no opportunity to compare simulated images with the original – there may be no possible perspective from which to view things like this, or it may even be that the depicted material does not exist in the real world. Hence, simulations may equip virtual worlds with visual and other qualities that do not mirror those of real-world processes. Ihde (this volume) analyzes the computer as a new 'epistemology engine' that succeeds the 'camera obscura' as the paradigm in epistemology.
3. Language is also an essential factor in the development of simulation. The evolution of complex and powerful programming languages has turned simulation into a manageable instrument. Algorithms implemented in software packages have made simulation methods, at least partly, a ready-made tool. The structure and features of programming language, for example, object orientation, determine to an important extent how programs can be conducted and how the practice, including the social practice, of programming operates. Shinn (this volume) considers the significance of this evolution in some detail.
4. Today, simulation has penetrated innumerable spheres of social experience, becoming manifest in ways totally undreamed of thirty or forty years ago! In the realm of medicine, 'artificial organ transplants' are tested in a simulated human body before being implanted in patients. Simulations form an essential part in the design and manufacture of technological artifacts from cars to bridges and buildings. The market for computer games and simulated film sequences is an instance in which increasingly more realistic virtual worlds are offered. What may reasonably be described as the cultural evolution of simulation, or co-evolution of culture and simulation, is also an important factor, because it opens up new atti-

tudes toward virtuality. Four decades ago, it was regarded as bad form for a theoretical physicist to rely on numerical computations, a professional methodological prejudice that is extremely rare today, as more and more scholars depend on the simulation technique. In sum, a growing number of people in contemporary society spend a significant part of their life in virtual worlds.

A coherent overall picture of the reciprocal relations between culture and simulation is still lacking, and this book does not aspire to a definitive view. Rather, the contributions gathered here try to clarify those questions and discuss some possible answers.

SIMULATION MODELING

Simulations are often portrayed as solutions to out-of-reach problems. Whereas this description is by and large valid, an important clarification has to be added. Problems are also often created and formulated as an 'answer' to new instrumental capacities. Thus, the new research instrument, and the problems and contexts this instrument is applied to, co-evolve. Some simulation approaches exploit novel strategies that formulate new problems, and apply entirely new kinds of models that were formerly unknown. We shall now consider some of these simulation models whose characteristics and particularities generate new practices and conceptions of modeling. But first we want to point to an important difference – the difference between numerical calculation and computer simulation.

Consider the calculation of a zero for a complicated function. The algorithm calculates the value of the function for two arguments so that the unknown zero lies between them. In a second step, one calculates the values at mesh points between the initial arguments. If a change in the sign of the function occurs between two mesh points, a zero lies in that interval of the mesh. Now one divides this interval into a finer grid, starts the calculation again, and so on. This rather straightforward procedure calculates zeros with arbitrary precision – a numerical calculation of a solution in the strict sense. Computer simulations, however, rely on procedures that differ fundamentally from numerical calculation in the strict sense. They do not simply solve complex systems of equations. Rather, simulations are numerical *imitations* of the unknown solution of differential equations, or, more precisely, the imitation of complex dynamics by a suitable generative mechanism.

The Monte Carlo method may illustrate our claim. This method is one of the first simulation methods dating back to the cooperation between mathematicians Stanislaw Ulam (1909–1984) and John von Neumann on the Manhattan project in Los Alamos. Imagine that one intends to determine the volume of a certain body via Monte Carlo – in most cases, an analytical integration is not possible. One can embed the body into a cube of which the volume is known. The idea is to replace the (unknown) primitive by a ratio that can be determined 'empirically,' or quasi-empirically, by iterating computer runs. The computer determines a point out of the cube at random. If this point belongs to the body, the trial is said to be successful. By repeatedly re-iterating this random choice, one can determine the unknown volume as the ratio of successful trials out of a great number of trials. In other words, the integration is imitated by a generative mechanism.

Ulam, surely one of the central figures in the development of simulation techniques, was enthusiastic about such 'statistical experiments,' appraising them as a new mathematical instrument. Von Neumann more reservedly looked at computational methods as a kind of emergency solution for when an elegant mathematical treatment is not possible. As mentioned above, even in the mid-1960s, to rely on the computer instead of working out analytical approaches as far as one could counted as bad form among theoreticians. Today, the situation has switched nearly completely. Simulations have lost their poor image and low ranking in the cognitive and epistemological hierarchy as a last resort, and have obtained a kind of autonomous status. They are clearly applied in many fields in which theoretical approaches do not even exist. The various simulation modeling approaches establish a spectrum in which the extreme cases start with a detailed theoretical model or, respectively, no model at all.

Consideration of the former case, in which one has a theoretical model at hand, suggests that even the most law-based simulation models may exploit the characteristics of behavioral imitation – instead of presenting a numerical solution of a theoretical model. Let us argue briefly for that point. In some branches of science, especially those related to physics, generally accepted theories provide laws that govern a system's dynamics. Typically, a set of coupled partial differential equations (PDEs) describe the relevant dynamic system. One of the main goals in the development of early computers was to design a machine that could tackle such systems of PDEs. Climate research presents an instructive example. It was initiated by John von Neumann, who considered the atmosphere's global dynamics as a paradigmatic out-of-reach problem (of hydro- and thermodynamics) that could be tackled by simulation methods. There exists a known and accepted set of PDEs, but the equations are too complex to be analyzed by traditional mathematical methods.

To mesh them with a digital computer, the continuous equations have to be transformed into discrete objects. Continuous *differential* equations are replaced by *difference* equations. Whereas the differential equations represent the functional (global) relationship between the variables, the difference equations determine the local values of the variables in different time steps. Starting with an arbitrary initial value, the variables are calculated step by step. If a specific grid in time and space is chosen, no values are available for the in-between variables. Hence, the continuous global processes described by the differential equations are replaced by a discrete series of local changes. The difference equation *imitates* the behavior of the continuous dynamics of the differential equation. Prior to simulations, modeling of complex dynamics, such as the atmosphere's global circulation, was commonly seen as impossible, lying far beyond the range of the available scientific tools. Even the existence of a simple model of PDEs able to describe the complex dynamics of the atmosphere was inconceivable. It was believed that the description of a complex system would require a model of at least the same complexity. The simulations approach changed that opinion. In 1955, Norman Phillips, who was then working in the meteorology group at Princeton's Institute for Advanced Studies, successfully conducted the so-called 'first experiment,' in which he used a set of five PDEs to model the complex dynamics of the atmosphere. He built a model of difference equations and managed to show by numerical experiment that they simulated important features of the atmosphere's dynamics. The reason for his success lies in the fact that complex systems may reach a steady state that depends on only a few variables and is resistant to perturbations from the external world. It is exactly that steady state – that is to say, a pattern of behavior – which is imitated by computer simulation.

After the first experiment had prepared the ground, simulation models became the focal point of atmospheric and climate research, shaping this discipline to a major extent. The question arises whether an observed steady state of a simulation is an artifact of the program or a realistic description of the dynamics under consideration. An answer to this question cannot be decided on the basis of the simulation itself. Therefore, simulations call for new ways of validating knowledge and assessing its uncertainties – in the case of climate policy institutions, a very practical task (see Petersen, this volume). Does the numerical treatment of the difference equation match the solution of the differential equation, that is, does the simulated atmospheric circulation match the dynamics of the theoretical model that is believed to be a realistic one?

This question can be answered in a strict sense only when analytical solutions are available. Without them, one has nothing to compare with. But such solutions are rarely at hand, and in most cases do not even exist. Thus, in general, the actual quality, or the goodness of fit of the simulation cannot be grasped mathematically! The difference equations approximate the differential equations when the grid becomes finer and finer. Ultimately, they become identical. In practice, however, one is bound to finite grid sizes, because the finer the grid, the greater the truncation error. Therefore, it remains unclear how adequately the difference equations represent the differential equations. In other words, there is no knowledge about the correspondence between the simulation and the theoretical model – it has to stand on its own grounds. Hence, extensive computational experimentation has to assure that the behavior of the system is simulated adequately and is not merely an artifact of the simulation method.

For this reason, computer simulations are not numerical *solutions* of a theoretical model: Rather, they employ a *generative mechanism to imitate the dynamic behavior of the underlying process.* Thus, starting from a highly theoretical law-driven approach, simulation practice 'cycles' to behavioral imitation! The result is a sound simulation, on the basis of differential equations, that imitates the dynamic behavior of the underlying process. Yet the issue of representational adequacy remains open (for a more detailed argumentation, see Küppers and Lenhard 2006).

This observation is even more true in the case of other types of model that do not rely on exact theoretical models – cellular automata, neural networks, and agent-based simulations. Cellular automata (CA) also employ a discrete generative mechanism to (re)produce the dynamic of a process. They divide the (often two-dimensional) state space into cells, thereby inducing a neighborhood structure of cells. Each cell can adopt different states, often two are sufficient; and it does so in each time step, dependent on the states of its neighboring cells. Simple rules for changing states can result in very complex dynamic patterns. It may be the case that a theoretical model serves as paradigm – Stephen Wolfram holds US patents for the treatment of differential equations via CA. Or it may not be the case – John H.

Conway's famous *Game of Life* (Gardner 1970) takes CA as autonomous, original generators.

Neural networks may be regarded as a similar case, perhaps marking the pole of the simulation models' spectrum opposite to the theoretical approach via PDE. Here, there is simulation of a system's behavior *without* employing the rules or laws that are thought to govern the real system's dynamics. It aims at a purely functional imitation. Such a network starts with a simple model of layers of neurons and then itself 'learns' or adapts the connection rules by optimizing its behavior. This learning process strives for functional or behavioral imitation, not for representation of the imitated system's structure. In the social sciences, so-called agent-based simulation models have been developed. In this context, agents are small autonomous programs that represent some knowledge and are able to communicate with others. These simulation models can reveal the complex behavior of social systems without using 'theories' of social interaction. For many years, artificial intelligence and other fields that investigate complex behavior focused on the reduction of that behavior to explicit rules. More recently, however, it has become clear that the context, that is, the boundary conditions of the dynamics, plays an equally important role in the emergence of complex behavioral patterns. Hauhs and Lange (this volume) argue in that direction when they draw some lessons from the failures when modeling ecosystems that are not only complex but also interactive systems. They argue that the inherent interactivity may be the fundamental obstacle to traditional modeling approaches, but that this may be dealt with by what they call 'interactive simulation.'

This observation of the often implicit properties of simulations is in no way disqualified by the fact that only 'explicit' commands constitute an algorithm. A complex dynamic may be based on few and simple assignments – albeit the resulting dynamics is not known in advance, but only from simulation experiments that render behavior visible. Again, consider a neural network, or CA: Admittedly, a program starts with explicit rules that specify the neurons and their synapses. The behavior of the network, however, is never characterized by the initial assignments! It is the very essence of this method that the simulation model does not incorporate the dynamic rules of the imitated system.

The neural network has to 'learn' and to adapt during the course of extended runs in which 'good' solutions become amplified and 'bad' ones are eliminated by positive or negative feedback loops. In other words, the simulation is a kind of numerical experiment in which different ways of operating are tested. This recursive process makes explicit some of the implicit properties of the model – relying fundamentally on this experimental strategy to achieve this goal. Numerical experimentation lies at the heart of a whole family of new experimental strategies and has been identified repeatedly as a key to the philosophical characterization of simulation modeling (Rohrlich 1991; Humphreys 1994; Keller 2003; Morgan 2003).

In complex systems, many simulation models may exist that produce the same dynamics. Therefore, very different generative mechanisms may be employed to tackle the same problem. A case in point is automatic translation. Automatic translation can be implemented in quite different ways: A structural approach would try to identify and apply the basic rules of grammar in the languages involved, thus setting a general semantic framework for the translation of concrete texts. In contrast, a functional approach is based solely on a statistical analysis of neighboring words and phrases in known texts that the computer first has to 'learn.' The translation of an unknown text is thus a kind of statistical estimate on the basis of the learned cases – not employing semantics at all! Both simulation approaches are being applied today.

Knuuttila (this volume) considers in detail the (functional) 'constrained grammar' approach to syntactic parsing that is a module needed for translation. This approach not only relies on a list of forbidden grammatical patterns that is implemented and can be processed and adapted effectively, but, at the same time, aims to imitate the behavior of a competent speaker. Different modeling approaches may even merge and give birth to hybrids like the so-called cellular neural networks (CNN) that are an effective technical – and US-patented – approach to deal with visual patterns (Chua and Roska 2002). Mathematically viewed, universal Turing machines, games of life, and cellular neural networks are equivalent in their ability to produce and transform patterns. They nonetheless present very different paths with respect to the technical implementation of simulations.

MEDIATION AND TENSION

Even a cursory reading of this book will document that discussion and debate about modeling play a strikingly prominent role. This observation reflects the significance of modeling practices for simulation, and, at the same time, the versatility of the current discourse about modeling in philosophy of science and in science and technology studies. The traditional philosophical account of models located them in the context of a theory – models are models of a theory. Today, this view has changed dramatically due to the beneficial influences of studies of scientific practice. A now famous controversy in philosophy was initiated by Nancy Cartwright's *How the Laws of Physics Lie* (1983) that attacked the assumed hegemony of theories and laws. Mary Morgan and Margaret Morrison's edited volume *Models as Mediators* (1999) advocates an account of models as 'autonomous mediators.' They thereby themselves play the role of a mediator between proponents of theory-based hierarchy and model-oriented pluralism. This discourse about autonomy and mediation effectively fits simulation and simulation models, because they mediate between theory, models, phenomena, and technology in new ways, highlighted by the investigation of simulation modeling practices (Winsberg 1999, 2003). In short, the new instrument of simulation offers new ways of mediating these elements.

The construction of models establishes an idealized model world that itself becomes an object of analysis. When leading to applications, this model world and the conclusions drawn from it have to be mediated with respect to the real world. Another case in which models have to mediate is between general laws and concrete applications. Recall the instance of atmospheric circulation in which a grand theory (hydrodynamics) had to be coined into a theoretical model, that is, a system of differential equations together with initial and boundary conditions. This model's behavior, which long lay out of reach, was brought into range by the superior power of the digital computer. We have argued that the theoretical model has to be modeled again to obtain a discrete simulation model that suits the abilities of a digital computer. Thus, simulations are kinds of model that additionally involve features of algorithmical implementation. The discrete simulation model imitating the atmosphere's dynamics has to mediate between theory, theoretical model, and phenomena, thus diversifying and further developing the autonomous mediating role. Winsberg (this volume) considers a simulation model from nanoscience that contains several autonomous components that become mediated by what he nicely calls a 'handshake' between autonomous regimes.

The mediation task becomes even more important when more attention is paid to the 'theory-free' pole of the simulation modeling spectrum: Simulations that imitate phenomena without representing underlying theoretical structures at all demand particular strategies of mediation. Simulations may be used to integrate a panoply of heterogeneous sciences as well as actors, scientists, and stakeholders. Climate science, for instance, embraces fields that range from physics and biology to economics and public policy. Simulations are placed in the midst – employed to forecast as well as to guide political negotiations. In what manner is simulation modeling organized to fulfill the various demands? Simulations are 'boundary objects' that impose mediating tasks, precisely because simulation lies on the level of institutions. Küppers and Lenhard (this volume) investigate how the organizational structure of climate research institutions and the architecture of simulation models interact.

If simulations imitate behavior – which criteria are appropriate to evaluate them? What makes a simulation a good simulation? That is, what are the validation criteria? In the case of real simulations like a flight simulator, it is relatively easy to decide on the quality of the simulation. Experienced pilots are able to judge with respect to their experience with real airplanes. In the case of *computer* simulations, there is little opportunity to compare a simulation with experience directly. In some cases, data from simulations can be checked with data from measurements. The computer model of the wings of an airplane and their dynamic behavior can be compared with 'real' data, but in the case of climate research, for example, the validation of predictive models is based mainly on the comparison with earlier time. The relevant data on the climate's history, however, are themselves reconstructed on the basis of models – a rather indirect relation. Petersen (this volume) considers the dangers of virtual worlds that may impress the experiencing subject because of their "speed, clarity, and internal consistency." The reliability of simulation-based knowledge, Peterson argues, poses new problems regarding how to assess virtual worlds appropriately and how to communicate with them effectively. The tension between reality and the virtuality of computer simulation, although sometimes acute, is often not exclusionary, leaving open possibilities for dialogue and for cycling back and forth between the simulation and the simulated. While frequently complex and difficult, this cycling proves fundamental to computer simulation's internal dynamics.

Alan Turing (1912–1954) situated the validation of a simulation in the setting of an "imitation game" (Turing 1950). An interrogator had to pose questions via telex and received responses from a human and a computer. Could he discriminate which set of answers was given by the machine and which by the human? A computer whose answers are indiscernible would imitate the answers of a human being and thus pass the 'Turing Test.' To date, no computer has succeeded!

Hence the imitation game that constitutes the Turing Test afforded a sophisticated arrangement to filter the functional equivalence out of the plenitude of theoretical possibilities. A computer has to behave like a human to pass the test, but the potential success has to be independent of the mechanisms that are employed to simulate. Turing, therefore, adhered strongly to a functionalist perspective. In short, Felix Krull passes a kind of 'Turing Test,' because he simulates the symptoms well. This standpoint has been challenged by a position that can be called 'structural' and that maintains that a valid simulation has to be structurally valid, that is, has to be based on a simulation model that represents the structure of the system under investigation. This problem has been taken up in J. Searle's (1980) famous and controversial 'Chinese room' argument that boils down to the point that imitation of intelligent behavior is not itself intelligent behavior. There have been a number of attempts to refine and redefine the Turing Test (Shieber 2004) based on controversial approaches to artificial intelligence: Should it provide a functional or structural imitation of cognition? For a long time, the computer – the electronic brain – was believed to be a structural model of the brain. Today it is known from brain research that cognition has nothing to do with calculation. The important point here is that the question "Is a computer imitating the behavior of a human being?" differs fundamentally from the question "Is a computer model an adequate model for human reasoning?" Artificial intelligence constitutes an early and important site for debate over this fundamental dilemma, but the tension between a functional (Question 1 about imitation) and a structural (Question 2 about models of mind) interpretation is of even broader significance for the simulation method in general.

In the preceding section, we discussed the example of automatic language translation. Structural approaches that try to implement the 'correct' rules of grammar compete with functional ones that focus on statistical analyses without modeling the grammatical structure of a language at all. Other cases are not that symmetrical. Economic predictions may be reliable if they are based on the 'correct laws' (for a critical discussion of this case, see Boumans, this volume). Climate simulations are held to be valid, because they are based on the physical laws that 'really' govern climate dynamics.

This structural viewpoint is attractive to practitioners of simulation, because of their attendant perception that it addresses the question of validity directly. Why should one trust simulation-based knowledge if successful imitation of behavior constitutes a consensually shaky category of evidence? Hence, the common view holds that a simulation is valid if it is based on a model that truly reflects the structure of the system under consideration. This form of argument is pivotal for many of those who adhere to the structural viewpoint. A popular benchmark in the technical literature is B. Zeigler's *Theory of Modeling and Simulation* (1976), in which he discusses different types of validity (replicative, predictive, and structural). Zeigler ranks 'structural validity' the highest, because it rests on the correspondence between the generative mechanism of the simulation and the real system, thereby offering an explanation of a simulation's successful performance. Consequently, according to the structural view, the aim is to implement the 'right' mechanism, whereas a functional perspective would be more liberal and cope with a potential plurality of mechanisms that may imitate a certain behavior under certain conditions. We seem to end up with a hierarchical order between structural and functional perspectives: The former is stronger, but more difficult to achieve. In many cases, a structural equivalence is far from evident, if not impossible to identify. The second choice would be the functional option: Reduce the demands on validity, don't be choosy regarding generative mechanisms, thereby permitting a gain in the number and range of applications.

However, to hold that a structural approach solves the problem of validation introduces a critical error in the appreciation of the performance of the category of elements referred to as 'models.' It is a rather general fact about models that they do not implement real structures or real generative mechanisms, regardless of whether they are structural or functional (whatever be their properties as models). Perhaps the hyperrealistic presentations employed by simulations lead practitioners of simulation to forget about the fundamental distinction between model on the one hand, and the modeled system on the other. Nevertheless, the problem of validation remains invariant regarding structural or functional approaches.

Recall the example of climate research in which the reliability of the simulations is said to depend strongly on the physical equations that form the basis of the simulation model. Cases like this are presented as the paradigms for the structural view, because the generative mechanisms of the simulation are said to coincide with the real ones. We have seen in the previous section on modeling, however, that even in this case, the modeling step from differential equations (representing laws of physics) to discrete, finite difference equations in an important respect makes simulation models a *functional* representation of the atmospheric dynamics. This transformation from the theoretical model to the simulation model is of utmost importance, because it changes the structure of the model. Hence, in most practical cases, the validity of the simulation model cannot be derived from the structural validity of the theoretical model but has to be judged by comparisons with the behavior of the real system.

This argument applies *a fortiori* to modeling approaches like cellular automata or neural networks. Hence, in general, at least some functional ingredients are inevitably part of simulations that imitate behavior by generative mechanisms. In fact, structural and functional approaches do not exclude one another, they are instead intertwined in simulation and set up a characteristic tension. The reader will encounter this tension repeatedly in this book. Boumans (this volume), for example, portrays how economic modeling has drawn its lessons from the Turing Test, and how the construction of an 'imitation' economy faces the tension between the functional and structural approach. He describes these standpoints aptly as white-box versus black-box modeling.

There is a similar situation with respect to scientific explanations. Humberto Maturana, a Chilean biologist and one of the founders of the constructivistic concept of autopoiesis, has also used the term 'generative mechanism,' not specifically in reference to simulation, but rather in reference to scientific explanations in general. He has suggested that explanations consist in "a reformulation of the experience (phenomenon) to be explained in the form of a generative mechanism" (Maturana 1990: 18), that is, a mechanism whose operation generates the phenomenon. Maturana, notwithstanding his radical constructivist account, suggests that an explanation would have to specify *the* generative mechanism that produces the phenomenon for which an explanation is sought. This places him in a structural framework. His account of validation thus encounters the same problem faced in simulation: The potential plurality of generative mechanisms seems to contradict a structural view that, in turn, offers the clue to valid results. In the case of heat, friction may count as a generative mechanism; chemical exogenic processes may count as another. It is sometimes possible to specify a basic generative mechanism for multiple physical expressions and thus provide a unified account. But one cannot expect this either in general or in practice. Thus, the tension between functional and structural accounts makes the question of validation an open one – and an intriguing one, particularly in simulation.

COMPUTER SIMULATION AS A SOCIOCOGNITIVE FIELD

The sociocognitive field of computer simulation is exceptionally complex and in constant flux, and is thus often difficult to characterize. Nonetheless, one can point to four persistent features:

1. Simulation practices are accompanied by considerable reflectivity. In view of the crucial character of cycling between the imitation/functionalist and the representational/structural perspectives, and with respect to the adoption and interplay of models in computer simulation and their persistence and omnipresence, this field is perhaps the most reflective domain of contemporary cognitive and technical life. The polysemy of computer simulations and the processes of cycling constantly call for an interrogation of the validity and limitations of one's orientation, the questioning of methodology on an extremely subtle and profound level, and the identification/establishment of the meaning of results and outcomes. Such also requires reflection by practitioners on the very grounds for their own reflection (Bourdieu 2001). Computer simulation does not just occasion efforts at objectivization of immediate activities and outcomes. It additionally invites analysis of the position of a practitioner with reference to alternatives, to the path adopted by others in the same technical and cognitive field, and, not least, it invites critical thought on one's own critical stances. In the absence of such reflectvity, the practices of computer simulation are at risk of quickly becoming either trivial or arbitrary to a degree far greater than in neighboring intellectual and technical domains.
2. Practitioners are engaged in a continuous re-embedding and reverse-flow re-embedding of methods, models, and perspectives in the course of movement back and forth between specific applications and involvement with simulation as a generic instrument. During re-embedding, the hub concepts expressed in a generic device, methodology, protocol, or language are transferred outward, and are adapted for adoption in a particular local environment with specific technical requirements. Reverse-flow re-embedding occurs when practitioners from local user niches combine their experienced-based indigenous practices with the surplus value derived from the acquisition of a generic device, and transfer the accumulated learning back to the generic apparatus. Johnson (this volume) gives an instructive example of this when she tracks down the movements of visualization experts and their knowledge to new environments, thereby concretizing the visual manifestations of nanoscience. Mattila (this volume) describes another example of a simulation model involving practitioners from mathematics, epidemiology, and health politics in which re-embedding processes evolve on different levels.

3. Due to the polysemy of computer simulation, and, for this reason, its innumerable domains of application, numerous computer simulators move across occupational and organizational boundaries. Multilayer meaning and multiple meaning characterize endeavors in computer simulation. This is connected to the centrality of the functional/structural issue and to the novel and complex dynamics of modeling in this field. It reinforces the opportunities for re-embedding and reverse-flow re-embedding just mentioned. Merz (this volume) discusses in detail the multilayered structure that simulations display in high energy physics – a whole chain of simulations is involved that connects tentative simulation experiments with the simulation-based design of large particle detectors. This chain is accompanied by various re-embedding and even dis-embedding processes of methods, objects, and researchers.
4. Computer simulators participate in the development and diffusion of a sort of simulation-linked 'lingua franca.' Metrologies, language, images, and competencies attached to hub concepts and methods in computer simulation become part and parcel of the procedures and skills of innumerable simulation user domains as simulation is taken up in them. The gradual growth of confidence in simulation and in its technologies, derived from application successes, comes to form a base for local validation and belief. Since the central precepts and concepts of computer simulation find expression in a similar manner wherever they are applied, the local practitioners and users of simulation evolve a common way of speaking, seeing, and doing – a lingua franca. It may be justified here to speak of the emergence of a kind of socially-based 'practical universality.' In his study of the Monte-Carlo simulation method, P. Galison (1996, 1997) coined the term 'trading zone' to designate that area in which boundary crossing between disciplines and actors takes place, and in which a common language, a 'creole' has to be developed to enable communication.

It has been suggested above that computer simulation is fundamentally an instrument; and it may indeed be effectively hypothesized that it is a 'generic instrument.' Simulations may be used as a kind of platform, like, for example, the SWARM simulation developed at the Santa Fe Institute that is intended to be a device to study macroscopic behavior based on individual dynamics as diverse as the behavior of fish swarms and the social dynamics of collective phenomena. A generic instrument may be defined as a device that incorporates and highlights a general instrumentation principle or concept. The principles are by essence open-ended, potentially allowing their expression in a large number of diverse domains.

Such genericity overlaps with our observed structure and operation of contemporary simulation. The *Science Citation Index* (SCI) database reveals some remarkable facts. The percentage of published papers with 'simulation' in the title relative to all articles in the database has grown linearly since 1960. It has reached about 0.6% which is not very astonishing, but the percentage of articles that mention simulation in keywords or abstract is growing much faster: It has doubled during the last decade, today attaining a level of more than 3% of total articles. This implies that the majority of papers that use simulation methods are not about simulation itself – nicely supporting our view of simulation as a generic instrument. Moreover, the disciplines that

SCI attaches to the journals that publish those simulation papers are extremely scattered. Physics, engineering, mathematics as well as the medical and even social sciences are discernible, but no concentration can be found. This finding again coincides with the universality of simulation thesis.

It is indeed the generic quality of the computer simulation instrument that has neutralized, in the case of simulation activities, the form of restriction and disciplining that are required of a body of learning in order for it to become an organizationally defined and closed academic discipline. Computer simulation is fundamentally heterogeneous with respect to practices, practitioners, and applications. Even the most cursory inspection of the occupational domains in which simulation is used and simulators are employed shows that there are hundreds and hundreds of them (see the list of members and tables of content of the annual meetings of the Summer and Winter Computer Simulation Conferences, Shinn, this volume).

A generic instrument possesses a kind of schizophrenic character: It functions in two spaces and two timeframes (Joerges and Shinn 2001; Shinn and Joerges 2002). The generic device incorporates, on the one hand, general simulation principles; on the other hand, the expression of said principles in concrete applications. In the one instance, there is a re-embedding of principles drawn from the generic apparatus in local situations; and, as will now be seen, there is also a reverse-flow re-embedding, as the experiences acquired during re-embedding are channeled back toward the initial generic device.

This process of re-embedding and reverse-flow re-embedding finds instantiation on two registers. The first instantiation pertains to the organizational dynamics of computer simulation. To design and develop a generic instrument, individuals often operate out of an interstitial environment located between the confines and requirements of dominant organizations like universities, firms, or state technical services (Joerges and Shinn 2001; Shinn and Joerges 2002). Whereas practitioners may be paid by one or another such body, they are nevertheless linked to it only loosely, and they often circulate freely between these bodies. He/she who works for everyone is the bondsman of no one. Simultaneously, re-embedding of generic principles in particular user domains demands computer simulation practitioners to cross occupational boundaries. They do this intermittently and selectively. Of utmost importance is that although the generic instrument may take root in an appropriately tailored form in a local space, it does not violate the specificity of the indigenous knowledge/practice base, nor does it violate the indigenous division of labor or organizational frame. Conversely, during the phase of reverse-flow re-embedding, in which local users inform the generic instrument, tactical boundary crossing of individuals away from their home coordinates takes place. Here, one discerns the operation of transverse social dynamics, a species of dynamics that connects and nourishes differentiated niches without imperiling their autonomy. Computer simulation can thus amply penetrate a host of application domains without jeopardizing them, and this constitutes one of the pillars of its universal success.

The second register of instantiation pertains to the homogenizing effects of the re-embedding and reverse-flow re-embedding processes associated with the dynamics of transversality. The presence in computer simulation of genericity, interstitiality, re-embeddings, and intermittent and selective boundary crossing has given rise to what might be termed fractal knowledge. At whatever segment of the whole, with whatever scale one looks (microscopic or macroscopic), one always discerns an underlying identical geometry. This disposition was perceived initially by Benoit Mandelbrot in pure mathematics and has since been observed in a variety of materials. The idea of fractals suits computer simulation in important respects. Through re-embeddings of the generic instrument of simulation in innumerable applications, and through reverse-flow re-embedding, there has come to co-exist a kernel concept of simulation and a range of heterogeneous expressions of that kernel. We have discussed a range of simulation modeling techniques and we have witnessed the characteristics of simulation-as-imitation on the coarse level of ordinary language, on the level of discussions in artificial intelligence or climate research, and again on the fine-grained level of difference equations that imitate differential equations. Stated differently, independent of whatever level one looks at, one discerns a basic computer simulation disposition – a kind of fractal simulation geometry. The efficacy of the generic principles in diverse environments, as directly experienced by practitioners and users, has produced a form of pragmatic proof of the robustness of simulation's validity. It has established the legitimacy of simulation as a concept and tool. This experience and the interactions between the generic device and local applications have led to the emergence of common skills and learning. This takes the form of shared techniques (the neural network model, cellular automata, Monte Carlo models, etc.), shared simulation-linked informatics languages (such as Simula or the general purpose, multiparadigm, object-oriented C++ simulation language, Shinn, this volume), shared competencies, shared images, and shared horizons. Taken together, such common resources, reactions, and perceptions constitute a lingua franca, itself akin in some ways to a form of practical universality. Such a practical universality, born mostly of communication spawned from a meaningful matrix of social and technical interaction, injects intelligibility into a professional, organizational, and intellectual/skill environment that is often characterized by extreme differentiation and even fragmentation. It may indeed be argued that computer simulation comprises one of the most pervasive and deeply rooted of all of the lingua franca that have arisen in recent decades.

Computer simulation has caused many aspects of science, culture, and society to change. The nature and scope of that change has not yet stabilized historically, and analysis of the dynamics remains an on-going endeavor. Indeed, the answers to a multitude of capital questions are not currently forthcoming; and it sometimes even proves difficult to identify the questions. No wonder, the instrument of simulation is still undergoing an evolutionary process, thereby transforming science, technology, and society – and still awaiting reflective considerations on multiple planes. As E.F. Keller has shrewdly and elegantly put it:

> Just as with all the other ways in which the computer has changed and continues to change our lives, so too, in the use of computer simulations – and probably in the very meaning of science – we can but dimly see, and certainly only begin to describe, the ways in which exploitation of and growing reliance on these opportunities changes our experience, our sciences, our very minds (Keller 2003: 201).

The contributions to this book identify, localize, and analyze some of the important changes that come with computer simulations. Foremost, they are taken to be an instrument for science and research that is applied in highly different ways and places – it is a *generic* instrument. Genericity, however, emerges out of the instruments' particular properties whose explication is the aim of this book. The term *Pragmatic Constructions of Reality* in its title alludes to the affinities of simulation to hyperrealistic models and experiences that not only represent the world but also create a new one: A virtual world.

From the very beginning, the computer was perceived as a *transparent* instrument capable of carrying out tedious tasks without sweat and without even leaving a fingerprint. Though simulations provide us with insights into the landscape of complexity, one must keep in mind a well-known principal distinction that may be expressed by the maxim 'the map is never the landscape.' Computer simulations can *imitate* the dynamics of a complex *process* or complex *function* by employing generative mechanisms. These mechanisms are construed in pragmatic ways and may employ sophisticated visualizing and experimental strategies. At first sight and on some levels, simulation even seems to overcome the above indicated ontological split between 'reality' and 'representation' with the aid of its simulation-generated, visually overwhelming images in the form of very landscape-like maps. But can they dissipate the fundamental tension between reality and its perfect imitations? The specific contribution of this book is to explore how computer simulations carry that relationship to the extreme. They do so in a way quite analogous to Gijsbrechts' 'simulated' painting that carries to the extreme the relation between illusion and representation.

ACKNOWLEDGMENT

We kindly acknowledge the financial support by the Volkswagen Foundation.

* *Bielefeld University, Bielefeld, Germany*
** *Maison des Sciences de l'Homme, Paris, France*

REFERENCES

Bourdieu, P. (1975). "La spécificité du champ scientifique et les conditions sociales du progrès de la raison", *Sociologie et Sociétés,* **7**: 91–118 (rééd.: (1976), "Le champ scientifique", *Actes de la Recherche en Sciences Sociales*, **2/3**: 88–103.

Bourdieu, P. (2001). *Science de la science et réflexivité,* Paris: Editions Raisons d'agir.

Cartwright, N. (1983). *How the Laws of Physics Lie,* Oxford, New York: Oxford University Press.

Chua, L.O. and T. Roska (2002) *Cellular Neural Networks and Visual Computing: Foundation and Applications*, Cambridge, UK: Cambridge University Press.

Galison, P. (1996)."Computer simulations and the trading zone", in P. Gallison and D.J. Stump (eds.), *The Disunity of Science: Boundaries, Contexts, and Power,* Stanford, CA: Stanford University Press, pp. 118–157.

Galison, P. (1997) *Image and Logic: A Material Culture of Microphysics,* Chicago and London: Chicago University Press.

Gardner, M. (1970). "Mathematical games: The fantastic combinations of John Conway's new solitaire game 'Life'", *Scientific American*, **223**: 120–123.

Humphreys, P. (ed.) (1994). "Numerical experimentation", in P. Humphreys (ed.), *Patrick Suppes: Scientific Philosopher*, Vol. 2, Dordrecht: Kluwer Academic Press, pp. 103–121.

Humphreys, P. (2004). *Extending Ourselves. Computational Science, Empiricism, and Scientific Me*thod, New York: Oxford University Press.

Joerges, B. and T. Shinn (eds.) (2001). *Instrumentation between Science, State and Industry,* Sociology of the Sciences Yearbook, vol. 22, Dordrecht: Kluwer Academic Publishers.

Keller, E.F. (2003). "Model simulation and 'computer experiments'", in H. Radder (ed.), *The Philosophy of Experimentation,* Pittsburgh, PA: University of Pittsburgh Press 2003, pp. 198-215.

Küppers, G. and J. Lenhard (2006). "Computersimulationen: Modellierungen zweiter Ordnung", *Journal for General Philosophy of Science,* submitted.

Maturana, H.R. (1990). "Science and daily life", in W. Krohn, G. Küppers and H. Nowotny (eds.), *Selforganization: Portrait of a Scientific Revolution*, Dordrecht: Kluwer Academic Publishers, pp. 12–35.

Mann, T. (1954). *Die Bekenntnisse des Hochstaplers Felix Krull*, Frankfurt a.M.: S. Fischer Verlag.

Morgan, M.S. (2003). "Experiments without material intervention. Model experiments, virtual experiments, and virtually experiments", in H. Radder (ed.), *The Philosophy of Scientific Experimentation*, Pittsburgh, PA: University of Pittsburgh Press, pp. 216–235.

Morgan, M.S. and M. Morrison (eds.), (1999). "Models as mediating instruments", in M.S. Morgan and M. Morrison (eds.), *Models as Mediators: Perspectives on Natural and Social Science*, Cambridge, UK: Cambridge University Press, pp. 10–37.

The Oxford English Dictionary, fourth edition, 1989.

Rohrlich, F. (1991). "Computer simulation in the physical sciences", in F. Forbes and L. Wessels (eds.), *PSA 1990*, vol. 2, East Lansing, MI: Philosophy of Science Association, pp. 507–518.

The *Science Citation Index* (SCI®), Thomson ISI.

Searle, J. (1980). "Minds, brains, and programs", *Behavioral and Brain Sciences,* **3**: 417–457.

Shieber, S. (ed.), (2004). *The Turing Test: Verbal Behavior as the Hallmark of Intelligence*, Cambridge, MA: MIT Press.

Shinn, T. and B. Joerges (2002). "The transverse science and technology culture: Dynamics and roles of research-technology", *Social Science Information,* **41:** 207–251.

Turing, A.M. (1950). "Computing machinery and intelligence", *Mind,* **LIX:** 433–460.

Winsberg, E. (1999). "Sactioning models: The epistemology of simulation", *Science in Context,* **12:** 275–292.

Winsberg, E. (2003). "Simulated experiments: Methodology for a virtual world", *Philosophy of Science,* **70:** 105–125.

Zeigler, B.P. (1976). *Theory of Modelling and Simulation,* Malabar, FL: Krieger Publishing.

IMITATING MODELS

CHAPTER 2

ANN JOHNSON

THE SHAPE OF MOLECULES TO COME

INTRODUCTION

Everyone knows that the origins of the computer lie at least partly in the development of scientific research in the twentieth century. The development of computing technology cannot be described without invoking key developments of twentieth century science, math, and engineering from the Manhattan Project to Bletchley Park to NASA. However, in most cases, the 'effect' of the computer on science is told with an assumption in mind – that adoption of the computer involved applying a new technology to already existing problems, theories, and methods. That is, science became computerized. Subsequently, the history of scientific computing often focuses on mutual adaptation – the ways scientific research and computing technology reciprocally transformed one another. Furthermore, there is clearly truth to this model – the *Electronic Numerical Integrator and Computer* (ENIAC) and the *Colossus* are neatly explained using this narrative. But, since the 1960s, this model has been increasingly outdated – the computer is a standard resource in the construction of new theories, experiments, and even new scientific disciplines – there is no longer a process of 'applying' the computer to long standing questions. Rather, the computer has changed the nature and form of the questions being asked. The new scientific questions that have been asked over the last third of a century are questions which would not even be asked without the existence of high-speed machine computation, and, increasingly, high resolution computer graphics. As a result, the adaptation approach, effective for the immediate post-World War II period, must be reexamined and supplanted by a new model – one which takes into account the dynamics of science and engineering that have developed during the computerized regime.

NANOTECHNOLOGY AS COMPUTER-AIDED SCIENCE AND ENGINEERING

Nanotechnology is a scientific and engineering research area that has developed since the early 1980s. Broadly speaking, researchers in nanotechnology work to show how molecules behave. But the emphasis in most nanotechnology is to create new types and combinations of molecules, which will yield new kinds of materials, and, even

J. Lenhard, G. Küppers, and T. Shinn (eds.), Simulation: Pragmatic Construction of Reality, 25–39.

more powerfully, will generate new structures built from these substances. But given its relatively recent development, nanotechnology's reliance on computing is total. That is, without computers, there is no nanotechnology. Unlike some disciplines in which there exists a continuity between present-day computer-aided research and the kinds of questions asked in the early twentieth century before the advent of scientific computing, computers, from supercomputers to PCs, predate nanotechnology. As a result, practices in nanotechnology have been built around computers. Of course, I do not suggest that computing technology is static. In fact, as computing capacity evolves, so do the uses of computational techniques and the kinds of questions that are computationally tractable.

Nanotechnology research uses computing technology in a variety of ways, which can be grouped into three broad categories of use. In experimental work, scanning tunneling and scanning probe microscopes use computers (essentially, CNC-technology) to control the movement of the probes. Then the data collected from the movement of the probe is fed into a computer, where software produces images of the surface of the substance being examined. All images of the nanoscale are generated using algorithms which map data into a graphical display. Pictures of the nanoscale are not photographic images; the images are produced through the meditating work of software engineers who write code to transform a series of data (i.e., measurements) into an image. Therefore, even what qualifies as the physical, experimental realm in nanotechnology research is heavily governed by computing techniques developed both within and imported into nanotechnology research.

However, there is another distinct category of computer use central to the construction of new knowledge of the nanoscale. A significant part of nanotechnology research takes place entirely on the computer, with only a limited connection to the physical world of the scanning microscope. This is the world of computer simulations of nanoscale structures (i.e., molecules). Since the 1990s, researchers have used computers to create a virtual, or simulated, experimental field. Simulated experiments can be performed on nanostructures, which, for a variety of reasons, are not possible in the physical world. This type of research in nanotechnology is called computational nanotechnology. This type of research is the focus of this paper, although, to reiterate, this is not the only or even the primary use of computers in nanotechnology. Computational nanotechnology is, however, the area in which the existing models of scientific knowledge production fall flat, stagnating in the muddy problem of differentiating theory, experiment, and model.

Given the centrality of computer-generated images to the practices of nanotechnology research, attention must be paid to the ways these images are generated. Instead of describing the role of simulations in nanotechnology in a way that accommodates older models of theory and experiment, here I argue that the way to understand the new role of simulations is to focus on the process of making computer models by asking questions such as the following: "What has been modeled thus far? Who has modeled it, and who employs the researchers? What are the short and longer term goals of the modeling projects?" The answers to questions like these can only come from specific case studies in computational nanotechnology. However, before discussing a specific case study, I would like to lay out a few more of the important questions related to the framework of this study.

Since the computer models of computational nanotechnology almost always result in images, it is important to define the nature of these pictures. Computer-generated images raise some of the same questions that photographic images do – principally, what does the image actually show? But because of their reliance on numerical data and algorithms, computer-generated images also raise unique questions, divisible into three epistemological categories:

1. Data – what data is being used as an input to the model? Where did it come from? What are its underlying assumptions?
2. Programming – What are the algorithms? Who created them? What assumptions do they carry with them?
3. Conclusions – What do the simulations show or prove? What kind of evidence do they provide for the expansion, modification, or refutation of some theoretical frame for the phenomena being studied?

These issues have both social and epistemological components. Understanding the processes by which scientific and engineering knowledge is produced often requires attention to the interrelated social and epistemological dynamics. By this, I mean the ways the research questions are informed by the particular individuals who ask them. Programmers ask different kinds of questions about simulation environments than do computational chemists, but both kinds of questions are part of the process of making new knowledge about simulated molecules. The kind of knowledge constructed depends on the kinds of people involved, and who is involved is conversely dependent on the kinds of research being undertaken. Social and epistemological dimensions are not neatly separable; neither category is prior to the other. With this framework in mind, I plan to focus on three questions to provide an entry into the broader questions about computer simulations that I lay out above: What is it that simulations of nanotechnology model? Who is using these models, in both institutional and personal terms? What are the researchers actually doing with the models?

COMPUTATIONAL NANOTECHNOLOGY – ORIGINS

The term 'computational nanotechnology' was first used publicly in a 1991 article in the journal *Nanotechnology,* by Ralph Merkle, a computer scientist, at that time working for Xerox PARC. Merkle drew both the term and its methods from computational chemistry. From the beginning, computational nanotechnology has depended on three computer-aided modeling methods from computational chemistry: molecular dynamics, semi-empirical methods, and *ab initio* methods. Merkle's claim for the utility of these methods in 1991 is that they would allow nanotechnologists to build and test molecular systems on the computer just as "Boeing might 'build' and 'fly' a new plane on a computer before actually manufacturing it" (Merkle 1991). Merkle's interest in computational or computer-generated models of molecular systems came from experimentalists' inability to build any of the systems he wanted to design. As a result, the computer screen acted as the only experimental space that would allow Merkle to design molecular machines.

But Merkle justified the use of computer-aided molecular design in a very specific way. He argued from his experience in computer science that computer-generated models in a number of fields had proven reliable indicators of the feasibility of various mechanical configurations. According to Merkle, knowing that a system will be feasible, even though the technology to actually construct it does not yet exist, will accelerate the process of developing the technology to build molecular machines. Computer simulations allow the development of assembler technology to be carried out in a parallel, modular process instead of a linear, serial one. In this way, computer simulations promise to accelerate the development of actual assembler technology by pinpointing the best routes to development. Merkle writes:

> Doing things in the simple and most obvious way often takes a lot longer than is needed. If we were to approach the design and construction of an assembler using the simple serial method, it would take a great deal longer than if we systematically attacked and simultaneously solved the problems that arise at all levels of the design at once and the same time. That is, by using methods similar to those used to design a modern computer, including intensive computational modeling of individual components and sub-systems, we can greatly shorten the time required to design and build complex molecular machines (Merkle 1991).

For Merkle, the computer allows technological development to proceed on a broad front. This approach makes an interesting contrast with Thomas Hughes' notion of reverse salients as catalysts for technological developments in Edison's lab. Hughes credits Edison with a great facility in seeing bottlenecks in the development of new technologies (Hughes 1983). Using language from military strategy, Hughes refers to these bottlenecks as reverse salients. Merkle hopes that computational models will help avoid reverse salients, yet Hughes shows that reverse salients play a crucial role in successful technological development and that their origins are diverse. One cannot predict, *a priori*, where the reverse salients will lie; they only become obvious once they surface. Therefore, the question about Merkle's desire to use computer models of nanosystems to allow development of many components simultaneously is whether the reverse salient problem will continue to exist. So far, most of the attacks about the feasibility have pointed to fundamental limits in technology to *make* nanosystems – as of yet, computer models have not eliminated Hughes' reverse salient problem. Furthermore, the reverse salient for a computational model may very well not be the one that presents itself in the lab or production process.

It is also important to mention that Merkle's claims and work echo that of Eric Drexler. Drexler's own work, especially that of the early 1990s, consisted of the kinds of computer models of nanosystems that Merkle was arguing would speed the development of actual, physical nanosystems (Drexler 1992). But Drexler was less dependent on established techniques in computational chemistry, drawing instead on computational models from physics and electrical engineering. Drexler also never referred to his work as computational nanotechnology, presumably out of concern that this qualifier would fuel his critics' arguments that his brand of nanotechnology was not actually possible (it was *only* computational). Merkle, as a computer scientist, used the term 'computational' with impunity – it was merely one kind of nanotechnological research, and a highly useful one at that. Merkle was also ready to claim a central role for computer science in the development of nanotechnology. In a 1999 article in *Technology Review*, speaking about self-replicating assemblers,

Merkle claimed: "Computer scientists are very comfortable with the idea. You can do it on a computer" (Voss 1999).

In the early 1990s, Merkle generated interest for computational nanotechnology in his own Silicon Valley backyard, and his work, at least in part, led to the creation of a computational nanotechnology research group in the NASA Advanced Supercomputing (NAS) Division at NASA's Ames Laboratory in Silicon Valley. There NASA has concentrated on developing techniques for modeling nanosystems, particularly novel nanomaterials and nanomachines that have not only extended computational chemistry techniques but also transformed these tools by adding other simulation methods from other disciplines. While Merkle claimed that the existing commercial computational chemistry software packages in 1991 were sufficient to design and test a number of critical components for the construction of molecular machines (on the computer, that is), the subsequent progression of new techniques has made that claim look increasingly naive (Musgrave et al. 1991). Over the decade of the 1990s, computational nanotechnology research has developed its own computer-aided methods that were more than simply borrowed techniques from chemistry, bringing in ideas, theories, software, and programming techniques from engineering and computer science.

NASA'S INTEREST IN COMPUTATIONAL NANOTECHNOLOGY

In 1998, Merkle co-authored a paper on "NASA applications of molecular Nanotechnology" in the *Journal of the British Interplanetary Society*. His six co-authors all worked at the NASA Ames Research Laboratory. The paper laid out a number of products and materials which NASA researchers were working on, all of which promised great importance in space research and travel. Most of the developments named in the article were to take advantage of significantly improved strength-to-weight ratios of nano-structured materials (Globus et al. 1998). It is in describing the novel properties of nanomaterials that computational nanotechnology has had the greatest impact. While the 1998 article was not exclusively devoted to computational methods, the importance of computer models was reinforced in the article's conclusion:

> … it is clear that computation will play a major role regardless of which approach … is ultimately successful. Computation has already played a major role in many advances in chemistry, SPM manipulation, and biochemistry. As we design and fabricate more complex atomically precise structures, modeling and computer-aided design will inevitably play a critical role. Not only is computation critical to all paths to nanotechnology, but for the most part the same or similar computational chemistry software and expertise supports all roads to molecular nanotechnology. Thus, even if NASA's computational molecular nanotechnology efforts should pursue an unproductive path, the expertise and capabilities can be quickly refocused on more promising avenues as they become apparent (Globus et al. 1998).

In 1997, the year before the appearance of this article, a research team for computational nanotechnology had been created at Ames. Locating the team at Ames was an important step, since Ames plays a critical role as NASA's primary computing research facility. The Ames location facilitated cross-fertilization between nanotechnology and developments in computer science and programming. In addition, computational nanotechnologists would be able to access easily and freely the very high-end computing facilities of NASA. Since the size of computational chemistry simulations were limited by computational power, and researchers thought these limits could be breeched only by faster computing power, this arrangement was crucial to the progress of computational nanotechnology.

The limits in computational chemistry came from the computational dynamics of scale versus detail – computational models of chemical systems ran along a spectrum with detail at one end (i.e., systems with quantum effects and temporal changes in bonding) and size (i.e., number of molecules involved) at the other. Computational methods had to balance this computational zero-sum game – examining either large systems with very limited detail or very detailed systems of very limited size. While classical molecular dynamics could be used to model stable systems of thousands of atoms, *ab initio* methods, which could deliver the energy states of more complex, unstable systems, were solvable for systems of less than 100 atoms. Computational chemistry methods in the mid-1990s were only feasible for molecular arrays of severely limited size and of a stable chemical nature – exactly the opposite of the kinds of novel nanostructures that organizations like NASA were interested in. *Ab initio* methods, which required the highest level of calculational complexity, were solvable for systems of less than 100 atoms. The molecular machines that Merkle envisioned modeling in 1991 contained hundreds of thousands, if not millions, of atoms.

Over a decade has passed since Merkle's original statements about the feasibility of computational chemistry tools, and yet the systems he was most interested in modeling have not been made fully tractable by developments in memory and parallel computing. Progress has instead come from modifications in simulation methods. Other techniques, from engineering, have been coupled with computational chemistry in order to simplify nanostructures and make their properties manageable. The last thing Merkle or NASA advocated was waiting for developments in computing technology to catch up with the kinds of problems they were interested in solving. Problems needed instead to be re-parsed in ways that would make them tractable. They found many resources for this process in the fact that mechanical engineers had been doing this kind of simplification for computing for decades.

CARBON NANOTUBES AND THE COMPUTATIONAL NANOTECHNOLOGY RESEARCH GROUP AT NASA-AMES

Work at Ames has focused largely on the carbon nanotube. Nanotubes are carbon molecules based on the Nobel prize-winning research on buckminsterfullerenes (a spherical arrangement of C_{60}) by Robert F. Curl, Harold W. Kroto, and Richard E. Smalley. In 1991 Sumio Iijima, of NEC's Fundamental Research Laboratories in Japan, discovered the nanotube arrangement of fullerene molecules. A single walled nanotube is essentially a rolled sheet of graphene made of six folded benzene type rings. Multiple wall nanotubes, which are easier to produce, can be thought of as a stack of graphene sheets rolled into a cylinder, capped with half-fullerenes. By 1992, Iijima had established that these molecules had some unique properties making them a very promising research focus. By 1993, Iijima's lab had produced single-walled nanotubes in very small quantities.

Three particular properties of nanotubes peaked NASA's interest (Srivastava 1997). First, nanotubes are potentially very high strength materials, several times stronger than diamond, but at the same time potentially less brittle. The magnitudes of nanotubes' strength and ductility still needed to be determined, both computationally and experimentally, but the underlying assumption at Ames was that nanotubes would offer significant advantages in their structural properties. Nanotubes could be used as both composite materials themselves and as threads in composites using other newly engineering materials – their unique strength-to-weight ratio, although yet to be precisely determined, recommended them for both. However, scale itself did pose questions (which continue to be debated), particularly whether the strength exhibited by a single molecule would translate to a comparable strength in the bulk material.

The second property that interested NASA was the electrical behavior of the nanotube. Single wall nanotubes could be manipulated to have the properties of either metallic electrical conductors or semiconductors, making them promising materials for both nanoelectronics and sensors (Hamada et al. 1992). They could even act as quantum conductors (Tans et al. 1997).

The third property, which was uniquely interesting to NASA, was the fact that nanotube arrays could, at least in computer simulations, be constructed into a hollow cage (Dillon 1997). The spaces between the atoms in the cage would be smaller than fuel molecules, which would allow fuel tanks to be constructed of this new molecule-scale mesh material. Since the weight of fuel tanks on spacecraft constituted a significant load, finding novel ways to contain the stored fuel would be one way to reduce the weight of space vehicles. Lighter spacecraft would be cheaper to launch, and reducing launch costs has been a NASA goal since the early 1970s, and was one of the driving motivations behind the whole space shuttle program (Coopersmith 2004).

Because of these three uses, the computational nanotechnology research program at Ames spanned NASA's four challenge areas (nanoelectronics and computing; optoelectronics; sensor technologies; and structural materials) and made computational work the keystone of nanotechnology research at Ames. The Ames web page on the computational nanotechnology research group defines their approach as follows, "Modeling and simulation across time and length scales coupling fundamental physics, chemistry, and materials science, and validation against experiments" (http://itp.arc.nasa.gov/computationalnano.html). As a result, simulations set priorities in laboratory work through the need, often unmet, to confirm or explain the results of simulated experiments.

The computational nanotechnology research team that developed at NASA-Ames was led by Deepak Srivastava, a well-established nanotechnology researcher with a PhD in theoretical physics. The nascent nanotechnology community has already recognized Srivastava with the 1997 Feynman Prize for Theory from the Foresight Institute. Srivastava focused NASA's nanotechnology research on an examination of the novel mechanical, electrical, and mechanical behaviors of carbon nanotubes. His models ran on the SGI SMP computer at Ames, as well as an SGI Origin 2000 supercomputer at the National Supercomputing Center. For Srivastava, a computer modeler from his graduate training on, simulations held the advantages Merkle had claimed – they were the only cost-effective way to actually design new nanomaterials. The shortcomings of computational models, based on the fact that they could only rarely be validated by physical experiments, promised to be erased by simultaneous developments in the laboratory.

Since 1997, Srivastava's work has focused on investigating a wide variety of mechanical, chemical, and electromagnetic properties of nanotubes. Unlike Merkle, who worked from methods already developed in computational chemistry, Srivastava wanted to determine parallels to macroscopic mechanical properties. To understand mechanical behavior, engineers begin with ordinary mechanical properties – elasticity, stress, strain, and others. Examining these properties at the nanoscale immediately brought new questions about the relationships between mechanical, chemical, and electromagnetic properties. In a macroscale structure, strain causes visible deformation of the structure, but rarely is there any concern about the strains having any effect on the chemical stability of the material – strain does not reverberate down to the molecular level in any profound way.[1] However, strain in a nanotube potentially deformed the chemical bonds holding the molecule together; as a result, strain could make the molecule chemically instable. That is, the strain could cause chemical bonds to be broken and remade in new locations. None of the simulation environments from computational chemistry were able to handle this kind of problem, since they did not model strain, and, for most, bonding is addressed only in *ab initio* simulations of very limited scale molecules.

However, from an engineering perspective, strain is easily defined in energy terms, which could easily bridge to the mathematical language of computational chemistry that also traffics in energy terms. As a result, new modeling tools have been created in order to simulate the chemical effects of mechanical strain. This was done by blending different techniques from computational chemistry as well as bringing in new approaches from mechanical engineering and materials science (Srivastava 1997). A new way of thinking about the behavior of molecules – new knowledge – was thus constructed, because a group of scientists and engineers came together from different disciplines, and existing assumptions became suspect to practitioners unfamiliar with them. Physicists and mechanical engineers called into question the scale assumptions of computational chemists, while computational chemists questioned engineers' and physicists' notions of physical strain and its potential effects on chemical bonding. These practitioners were also used to working at different scales – crudely put, physicists concentrate on the quantum scale, chemists at the atomic and molecular, and engineers at the molecular or bulk. The result of this interdisciplinary collaboration was a set of new simulation techniques that produced solutions to particular questions the researchers were asking about previously intractable molecular systems.

Srivastava's work at Ames is best characterized as a hybrid, according to the usage of the term in Peter Galison's *Image and Logic*. According to Galison, hybrid practices are detectable in both the social makeup of research teams, looking at where the members have come from and the ways they are arranged in the research effort. But hybrid practices are also evident in the artifacts and processes that the researchers construct. The seams between various methods are often visible – especially at the beginning – and in physical artifacts, the hybridity of the components is also apparent (Galison 1997). In the case of Ames computational nanotechnology group, members of the group came from physics, chemistry, mechanical and chemical engineering, and computer science. As Galison predicts, the hybridity of the group is reflected in the approach and the modeling tools they produce. Beginning with basic computational chemistry methods, the Ames group began with a simulation of single-wall carbon nanotubes. Using existing molecular dynamics tools, they wanted to show the elastic instability of the nanotube. In other words, when the tube was put into compression, it remained straight until a threshold was reached, at which point it buckled. However, there was little elastic deformation prior to the buckling. At this buckling point, the chemical bonds of the tube became unstable and the tube was permanently (plastically) deformed. In 1997, Srivastava wanted to extend this process to multiple-walled nanotubes, which he hoped would be stiffer and less prone to this catastrophic kind of deformation. Multi-walled nanotubes presented much more difficult problems of analysis. They were comprised of a much larger number of atoms, which was a critical problem for computing capacity. Second, they required computational tools which would describe changes in long range van der Waals interactions. Ignoring these meant assuming a multi-wall nanotube acted the same as a single wall nanotube, which was both counterintuitive and contraindicated by the limited laboratory experiments on nanotubes in 1997.

RESULTS OF THE NANOTUBE MODELS AT NASA-AMES

The Ames group began with a FORTRAN77 code written in 1990 by D. W. Brenner, a materials scientist with a PhD in chemistry. The Brenner-Tersoff potential approximates the potential carbon-carbon interactions of a complex molecule, by using the Hamiltonian of Newton's Second Law to define potential energy functions in an array of atoms. The unique aspect of the Brenner-Tersoff potential is that it allows short-range bonds to be interactive, meaning that bonds can be made and broken in the course of the simulation. This feature also means that the simulation is much more costly in computational terms than a nonreactive simulation. However, the Ames group also had access to a SGI Origin 2000 supercomputer at the National Center for Supercomputing, which allowed them to be less concerned about computational demands. Previously, they had used SGI's SMP, a four processor system. The Origin 2000 was a distributed shared memory system, which could run the previous code, but also had a great deal more capacity. Code written for the SMP was optimized by the Ames group for use on the Origin 2000, allowing them to benefit maximally from the supercomputer's power, running simulations of a much larger number of atoms than the SMP could handle. Using the Origin's parallel shared memory architecture, the Ames group simulated a system of over 100,000 atoms. This scale of investigation required changes in Brenner's code, since his focus was to look in great detail at a small system. The real challenge facing the NASA group in examining multi-walled nanotubes was that they wanted to see tight detail, dynamic processes, and a large number of atoms – all of which demanded intense computation. This kind of complexity had not been modeled previously.

Still, the simulation was created with an outcome already in mind – they were seeking to show that the multi-walled nanotubes would present a solution to the catastrophic buckling problems that they had already simulated in simpler single-walled nanotubes. And the simulations demonstrated these assumptions to be true, at least in most cases. The multi-walled tubes buckled less because the inter-wall van der Waals forces prevented the worst pinching and deformation. For the kind of severe deformation seen in the single-wall nanotubes to occur, the walls of tube needed to be closer together than the relatively weak van der Waals normally allow (3.4Å). Still, as strain increased in the multi-walled tubes to 17%, the deformation became plastic and the tubes came apart, as had the single-walled tubes at a lower strain level (8.5%). But multi-walled nanotubes appeared in these simulations to have significant benefits over single-walled tubes. In the range of 5-16% axial strain, the strain curve was linear and largely elastic. Only at the extreme end of 17% strain did the sudden transformation to plastic deformation occur. Given the high level of strain required for this transition, double what was required to deform single-walled nanotubes, multi-walled nanotubes looked like an improvement. Furthermore, multi-walled nanotubes are more easily and reliably produced. This simulation was considered a great success and led to further funding of the computational nanotechnology group at Ames.

The Ames group also ran comparable simulations on single-walled nanotubes, even though, because they were only single-walled, they would not have the advantage of the van der Waals forces between the layers of graphene to strengthen them against buckling. In a way, this repetition of earlier simulations using a new method was a way of checking the simulations against previous simulations, the only verification possible, since experimental tests were not feasible. While previous simulations of strain in single-walled nanotubes had been of tubes 7 nanometers long, the Ames group ran tests on 45 nanometer tubes as well and found that they also had unique properties owing to their length. The longer tube simulations showed the researchers that the deformations were highly localized and at random locations – showing that the results of the shorter tubes could not simply be scaled up for the longer ones. The strain appeared to accumulate at the 'sharp kink' locations and bonds were rearranged at those locations – being broken by the geometry of the kink in some locations and rebonded in others. The group called for a similar set of experiments on actual nanotubes using an atomic force microscope to confirm the veracity of the simulations. However, in 1997, the production of nanotubes was not reliable enough to facilitate this type of work.

SECOND GENERATION SIMULATIONS AT AMES

By 2000, the methods used to model strain in nanotubes had developed significantly more quickly than had laboratory techniques to confirm the results of simulations. Consequently, a decade after Merkle's claims about the importance of computer simulations in speeding the development of nanotechnology, it was ironic that the computer simulations that were supposed to speed development had so far outstripped the development of physical nanosystems. It would seem that the real catalytic effect of computational nanotechnology was on computational nanotechnology itself. Still, in a 2001 article in the journal *Computing in Science and Engineering*, Srivastava restated Merkle's words from a decade earlier:

> The role of *computational* nanotechnology has become critically important in nanotechnology development. The length and time scales of nanoscale systems and phenomenon [sic] have shrunk to where we can directly address them with computer simulations and theoretical modeling with high accuracy. The rapidly increasing computing power used for large-scale and high-fidelity simulations make it increasingly possible for nanoscale simulations to be also predictive (Srivastava et al. 2000).

While Srivastava clearly agreed with Merkle's devotion to computational nanotechnology, it is worth noting that Srivastava did not make claims about the importance of computational nanotechnology as an accelerant to the development of nanosystems. Furthermore, there is a subtle shift from thinking in a scientific, theory-based, 'knowing that' manner in Merkle to an engineering, prediction-oriented, 'knowing how' attitude in Srivastava. This shift is also reflected in Srivastava's collaborators for the 2001 article – Madhu Menon, the associate director of the Center for Computational Sciences at the University of Kentucky, and Kyeongjae Cho in the mechanical engineering department at Stanford. In the 1990s, computational nanotechnology had taken the existing tools of computational chemistry and used them in new ways to solve new problems, which was possible through marrying them with techniques from mechanical engineering. This new way of working is evident in Srivastava's and the Ames' groups' work at the turn of the twenty-first century.

The Ames group was still using molecular dynamics models to describe the interactions of the atoms in the system, but they were also using a number of other tools. The most significant change between 1997 and 2001 was the adoption of a common technique from engineering, which was to vary the scale of examination by using different methods to model different parts of a structure. This approach had been developed during the initial phase of computerization in engineering in the 1950s and 1960s (Johnson 2004). An engineer designing a structure would model areas with possible stress concentrations in a highly detailed manner, while using much looser modeling techniques for areas where critical stresses were not expected. This allowed computer time and capacity to be used where they were most important and not wasted on areas where problems were not anticipated. Some engineers, and more commonly physicists, fought against this approach, relying as it does on the modeler's intuition and experience in determining a priori where problems lie. But despite some resistance, this technique has become a mainstay in engineering computer-aided design. By 2001, Srivastava was using this idea to model nanosystems, looking at varying levels of detail by using different tools to model different parts of the nanostructure. Eric Winsberg's piece in this volume examines in detail this process in a different case study.

MULTIPLE TOOLS FOR MULTIPLE SCALES OF EXAMINATION

The Ames group divides their methods into three categories drawn from computational chemistry: classical molecular dynamics methods, semi-empirical quantum simulations, and *ab initio* quantum molecular dynamics and density functional theory (DFT) methods. The classical molecular dynamics methods are similar to the Brenner-Tersoff modeling techniques used in the 1997 work to show the effect of the van der Waals forces on the elastic/plastic deformation of a multi-walled nanotube under compression. Even simpler molecular dynamics models are sometimes used, which ignore the chemical bonds that may be formed and broken during the event simulated. For example, in single-wall nanotubes, where there are no van der Waals forces between layers (because there is only one layer of graphene), classical molecular dynamics models optimize computing effort by eliminating bond interactions that do not exist, or in other cases, are not relevant.

But in many nanoscale systems, quantum effects start to matter, and classical MD methods are not effective in these cases. In moving to a quantum approach called tight binding molecular dynamics (TBMD), the Ames group modeled the atoms as a collection of quantum mechanical particles governed by the Schrödinger equation. In this approximation, the electron is tightly bound to its own atom except during "the capture interval, when the electron can orbit around a single ion uninfluenced by other atoms, so that its state function is essentially that of an atomic orbital" (Srivastava et al. 2000). The advantages of TBMD are its computational efficiency and the method's facility to determine the energy potential of a nanosystem without excessive computing resources. The development of more elaborate TBMD models has required the construction of new algorithms, and this has interested the computational geometry community.

At the extremely computationally demanding end of the spectrum of modeling tools used at Ames are *ab initio* simulation methods. *Ab initio* simulations solve complex quantum many-body Schrödinger equations with numerical algorithms. The *ab initio* methods are approximately a tousand times less computationally efficient than TBMD methods (Srivastava et al. 2000). As a result, *ab initio* models are limited to systems no larger than a few hundred atoms. The Ames group uses *ab initio* methods in conjunction with TBMD in order to vary the level of detail they need on atomic scale processes. The most important *ab initio* method for Ames is driven by density functional theory, which shows that the ground state total electronic energy of a many electron system is a function of the system's total electron density. At Ames, they use a specific algorithm called the Kohn-Sham equation, a single electron Schrödinger equation that can predict material properties "without using any experimental inputs, other than the identity of the constituent atoms" (Srivastava et al. 2000). This method is available through a commercial DFT simulation program, the *Vienna Ab Initio Simulation Package*. Despite the computational inefficiency of DFT methods relative to semi-empirical or classical MD, there have been significant improvements in recent years, which have allowed larger and larger systems to be modeled. However, it also seems likely that the modeling projects at Ames will continue to be composed of multiple simulation tools in order to model precise effects in the most efficient manner. Since the end goal of many nanosystems is to create macroscopic effects or even artifacts, there is no boundary on the size of systems to be modeled, therefore figuring out computationally efficient ways to predict energy levels promises to be a continuing focus in computational nanotechnology.

The way these three groups of modeling techniques have been used at Ames is illustrative of the practice of finding particular locations, energy levels, or processes that require greater detail and sharper focus in the modeling phase. When simulating carbon nanotubes to determine their mechanical characteristics, researchers at Ames first ran classical MD models to determine the limits of elasticity, as previously described in the 1997 work. When they found nonlinear elastic instabilities (which they called 'fins') in the deformed tubes, they ran TBMD simulations to further investigate the deformities. With TBMD models, their results were similar to classical MD models for strain values less than 8% (this was also the strain level where differences were seen between single wall and multi-wall nanotubes). But with strains greater than 8%, changes in the bonds occur and cause the molecular structure of the nanotubes to collapse inward. As new diamondoid bonds form, the structure is pulled even further into itself. In addition to running TBMD simulations of carbon nanotubes, they also ran the same simulations on Boron-Nitride (B-N) nanotubes. These results were also different for B-N nanotubes, since the B-B and N-N bonds make these tubes less stable than the C-C bonds. Changing the geometry of the nanotube from structures that look like _/‾_/ (called an armchair nanotube) to ones with bonds that look like /\/\/\ (called a zig zag nanotube) was one way to overcome the instability of the B-N nanotubes, and work in this direction was a direct outcome of the computer simulations. Srivastava and his collaborators claimed in their article that "this set of simulations for nanotube mechanics shows that in the nanoworld, simulations not only an verify and explain experimental observation – they can also predict new phenomena" (Srivastava et al. 2000) But it is clear that in the three to four years since the creation of the computational nanotechnology group at Ames, the researchers have drawn in many different kinds of resources, all of which moved their work toward a predictive, engineering design direction. Computational nanotechnology at Ames uses many tools from computational chemistry, but in conjunction with tools from computer science and mechanical engineering. Furthermore the ethos, or way of thinking, of the Ames researchers seems to have more in common with engineering design than theoretical chemistry. In this sense, computational nanotechnology is clearly a hybrid field.

WHERE DOES COMPUTATIONAL NANOTECHNOLOGY LEAD?

Computer simulations of nanoscale objects moved from being an application of computational chemistry in the early 1990s to a well-developed research area in the exploding field of nanotechnology by the turn of the century. One of the difficulties of discussing the evolution of computational nanotechnology is the simultaneous development of nanotechnology. It may be hard to see computational nanotechnology as a subfield, because some would argue that nanotechnology does not yet rate as a discipline. Yet, for the purposes of examining computer simulation, this argument seems peripheral. Computer simulation has been central to the development of nanotechnology – to the extent that it has been and remains the leading edge of the field. Partially due to the difficulties presented by experimental nanotechnology, computational nanotechnology is the validation test field for much of nanotechnology. Still, this position presents problems exactly because simulations are neither physical experiments nor first-principle-based mathematical models. Srivastava and other Ames researchers do claim that simulations act as feasibility proofs for nanoscale systems – they show what could be possible given our current understanding of the physics and chemistry of molecules. In this sense, and Srivastava makes this claim, computational nanotechnology has taken a role very similar to finite element models for engineering design – they show what is possible, what design aspects require further investigation, and how different techniques can be combined in highly contingent ways to maximize computational efficiency and minimize design throughput time. Computational models play a key role in ruling out certain options and concentrating researchers' efforts on areas which simulate positively. But the similarity of computational nanotechnology models to finite element models for macrostructures also shows the potential problems of computational methods, especially when based on thin experimental data. In the nearly half century since the development of finite element analysis, numerous examples of computer-aided failures have accumulated (Petroski 1999). In macroscale engineering, failures fly in the face of FEA models, requiring reexamination of the models as well as the failed structures. In a peculiar sense, this is a luxury that computational nanotechnology does not yet have. In nanoscale engineering, the line between technical capacity and physical feasibility remains unclear, and this is a distinction that even the most complex simulations cannot fully clarify.

University of South Carolina, Columbia, SC, USA

NOTES

[1] However, fracture mechanics is an area in which the relationships between mechanical stress and molecular bonds are critical. In addition, heat-induced strain can be chemically significant, as heat itself causes bulk deformation as well as changing a material's properties. Nowhere was this more graphically illustrated than in the collapse of the World Trade Center Towers. However, the concerns NASA was investigating were neither fracture- nor heat-related.

REFERENCES

Coopersmith, J. (2004). "No cheap ticket to outer space", on *History News Service* website http://www.h-net.msu.edu/~hns/articles/2004/020104a.html.

Dillon, A., K. Jones, T. Bekkendahl, C. Kiang, D. Bethune, and M. Heben (1997). "Storage of hydrogen in single-walled carbon nanotubes", *Nature*, **386**: 377–380.

Drexler, K.E. (1992). Nanosystems: Molecular Machinery, Manufacturing, and Computation, New York: John Wiley & Sons.

Galison, P. (1997). *Image and Logic*, Chicago, IL: University of Chicago Press.

Globus, A., D. Bailey, J. Han, R. Jaffe, C. Levit, R. Merkle, and D. Srivastava (1998). "NASA applications of molecular nanotechnology", *The Journal of the British Interplanetary Society*, **51**: 145–152.

Hamada, N., S. Sawada, and A. Oshiyama (1992). "New one-dimensional conductors: Graphitic microtubules", *Physical Review Letters*, **68:** 1580–1585.

Hughes, T.P. (1983). *Networks of Power*, Baltimore: Johns Hopkins University Press.

Johnson, A. (2004). "From Berkeley to Boeing: Civil engineers, the cold war, and the origins of finite element analysis", in M.N. Wise (ed.), *Growing Explanations: Historical Pespectives on the Sciences of Complexity*, Durham, NC: Duke University Press, pp. 133–158.

Merkle R. (1991). "Computational nanotechnology," *Nanotechnology*, **2**: 135.

Musgrave, C., J. Perry, R. Merkle, and W. Goddard (1991). "Theoretical studies of a hydrogen abstraction tool for nanotechnology", *Nanotechnology*, **2**: 187–195.

Petroski, H. (1999). *Remaking the World*, New York: Vintage Books.

Srivastava, D. (1997). "Molecular dynamics simulation of large scale carbon nanotubes on a shared memory architecture", in *Proceedings of IEEE Supercomputing*, Los Alamitos: IEEE Computer Society Press, p. 35.

Srivastava, D., M. Menon, and K. Cho (2001). "Computational nanotechnology with carbon nanotubes and fullerenes", *Computing in Science and Engineering*, **3** (4): 42–55.
Tans, S.J., M. Devoret, H. Dai, A. Thess, R. Smalley, L. Geerligs, and C. Dekker (1997). "Individual single-wall carbon nanotubes as quantum wires", *Nature*, **386:** 474–479.
Voss, D. (1999). "Moses of the nanoworld", *Technology Review*, **102** (2): 60.

CHAPTER 3

TARJA KNUUTTILA

FROM REPRESENTATION TO PRODUCTION: PARSERS AND PARSING IN LANGUAGE TECHNOLOGY

INTRODUCTION

In many interesting case studies, social research on science has demonstrated how facts and new scientific objects are created, or *constructed*, in scientific practice. The findings of these studies have challenged the representationalist view of science, according to which knowledge consists of a bundle of accurate representations of an independently existing external world. In opposition to this view, an alternative *performative* conception has been presented that conceives of science as a production process in which the artificial and the real are intermingled (see, e.g., Hacking 1983; Pickering 1995). The performative conception of science works well in the case of experimentation: It does not lead to any exaggerated form of relativism, as the opponents of 'social constructivism' once feared. Indeed, laboratory work, which has been the main focus of constructivist science studies so far, seems well grounded in physical reality, as its experimental effects are produced with the help of the material machinery.

When it comes to modeling, however, this neat distinction between performative and representational conceptions of science becomes muddled, and nagging questions concerning representation persist. This is because models are typically made out of different stuff and embodied in a different scale than the things modeled. Whereas experimentation under laboratory conditions means experimenting with the objects of interest themselves, the experiments provided by models seem to be virtual at best, giving us merely 'surrogates' for reasoning. Consequently, philosophers have tended to analyze the epistemic value of models more or less solely in terms of representation. Irrespective of the other differences in their philosophical outlook, philosophers of science have been nearly unanimous in holding that if they are to give us knowledge, models have to be representative (e.g., Morrison and Morgan 1999; French 2003; Giere 2004). Often, an even stronger claim has been made: Models *are* representations (e.g., Hughes 1997; Teller 2001; Frigg 2003). However, this way of ascribing the epistemic value of models to their representative aspects then tends to overlook the epistemic importance of many more performative aspects of models:

J. Lenhard, G. Küppers, and T. Shinn (eds.), Simulation: Pragmatic Construction of Reality, 41–55.

They typically function also as heuristic tools, demonstrations, test beds, and as detachable 'templates' (Humphreys 2002) that can be applied across the disciplines.

This paradoxical nature of modeling is especially striking in the case of simulation models, which, from the very outset, seem to be "caught between machine life and symbol life" (Galison 1996: 139).[1] This is already reflected in the way the early practitioners used to talk about simulations as *theoretical experiments* (p. 138). Nevertheless, even when the simulation consists of a 'discretization' of a theoretically well-established differential equation, the resulting simulation model is not a direct transformation of the original theoretical formulation (e.g., Winsberg 1999, 2003). What is more, the approximations, idealizations, and even falsifications typical of modeling are, in the case of simulation, importantly tied to the affordances and constraints of a specific material machine, namely, the computer. Thus, what Galison calls "symbol life" is configured and prepared to fit the machine life that makes it experimentable in the first place. Moreover, simulations are typically valued for their functional properties, whereas less stress is placed on their being structurally representative.[2] On the other hand, one epistemically important dimension of simulations consists in the *output representations* they generate (see Humphreys 2004). These properties of simulations call for new analyses that investigate how the representational and performative aspects relate to each other in modeling in general and in simulation in particular.

In the following, I shall attempt to give a new account of models that is better able to capture the different epistemic characteristics of models and modeling. I proceed by studying the Constraint Grammar Parser, which can be characterized as a computational model of syntax, whose task is to give the same kind of linguistic analyses of a written text as a human (linguist) can. Thus, the Constraint Grammar Parser is a simulation in the sense that it is both a stand-in for a human linguist and a computer program that executes the grammatical constraints of a specially devised language model. Instead of asking whether or not simulation constitutes a new mode of doing science that presents us with a radically new kind of tool, I rather take it that simulation models give us good cause to reflect on the interrelationships between productive and representative aspects of modeling.

I shall specifically argue that the traditional emphasis on representation does not give due consideration to the epistemic intricacies of model building and use, in which a special kind of expertise, bound to the specific models and methods of modeling, is seen to emerge. This seems to be especially true of simulation models, which, in addition to being models, are complex technological artifacts. Yet representation *has* an important place in modeling, though not the one it has been granted most often – as is particularly well demonstrated by computer models. First, often in the case of simulation models, the epistemically interesting locus of representation has moved from the structure of the model to its output. Parsers, for instance, are valued for what they *produce* and the accuracy of their output, rather than for being realistic representations of human linguistic competence. Second, even though parsers cannot be considered as straightforwardly representative entities, they are nonetheless constructed by various representational procedures. Consequently, the case of parsing challenges our established views on representation. It shows that much of the *work of representation* that takes place in modeling remains invisible if we are mainly interested in what the ready-made models supposedly stand for in reality. This, in turn, leads us to realize that what is being represented when we build models is usually something that has already been theoretically rendered and conveyed somehow.

CONSTRAINT GRAMMAR PARSER[3]

The process of describing a word or a sentence grammatically is called *parsing*. It has long roots in Western grammatical tradition with origins in the Latin *pars orationis*, which means 'part of speech.' A part of speech is a class of words with shared grammatical characteristics that distinguish them from other words. Generally recognised parts of speech are, for example, nouns, verbs, pronouns, adverbs, adjectives and conjunctions. In the context of language technology, parsing means "the automatic assignment of morphological and syntactic structure (but not semantic interpretation) to written input texts of any length and complexity" (Karlsson1995: 1). A parser is thus a language-technological device, a program devised for producing the parsed text necessary for various different language-technological tasks such as word processing, grammar-checking, and information retrieval.

There are two primary approaches to parsing. One is a grammar-based, linguistic, and descriptive approach; the other is probabilistic, data-driven, and statistical. The methods of the data-driven approach include corpus-based learning rules, hidden Markov models, and machine-learning approaches. Although the probabilistic approach is more popular, good results have been achieved with both methods. I shall concentrate on the grammar-based approach, that is, on the *Constraint Grammar Parser for English* (EngCG), which is a product of long-term research in language technology being carried out at the Department of General Linguistics at the University of Helsinki. Even though EngCG is a grammar-based parser, it is also firmly grounded on linguistic corpora. Thus, the constraint grammar approach to parsing differs radically from Noam Chomsky's universal grammar (1957), which is generative in the sense that it strives to define a set of rules capable of generating all well-formed, that is, grammatical, sentences of the language. Now, this idea seems to be closely analogous to the structural approach[4] to simulating physical systems in its aim to generate the complex phenomena of language with the help of relatively small set of simple rules applied to primitive atomic sentences.[5] Constraint Grammar (CG), in contrast, remains entirely on the level of surface structure, and instead of stipulating the rules for well-formed expressions, it consists of constraints that reject the improper ones. I shall come back to the distinctive features of the different approaches to parsing later in this article.

Constraint Grammar is a parsing formalism whose main task is to give a correct grammatical reading to each word of a running text and to enrich each word with further syntactical information. CG parsing builds on a preceding morphological analysis performed by a so-called morphological analyzer. Such an analyzer is two-level morphology (TWOL, Koskenniemi 1983), whose development was the basis for the Helsinki group's early international breakthrough. TWOL strives to give *all* the possible morpho-syntactic readings for each word of the text to be analyzed. Typically, the words we use are such that the same word form, say 'round,' can be read differently depending on the context in which it is used. Thus, TWOL gives the word ‘round’ a total of eight different readings, of which four are verbal; however, it can be read also as a preposition, noun, adjective, or adverb (Voutilainen 1995: 165). Consequently, the word ‘round’ is *ambiguous*; without any additional contextual information, we cannot decide how to interpret it. The task of the parser is to choose which one of these readings is the proper one in the context of its occurrence in the text. This is called *disambiguation*.[6]

Parsing proceeds as follows: Once the morphological analyzer has provided all the correct morphological readings of the words of an input text, the parser checks which readings are appropriate by applying morphological *constraints*. The constraints make use of the context of each word (i.e., the words in the vicinity of the word in question), whereby the clause boundaries provide the limit for the relevant context. For instance, if a word has both nominal and verbal readings and is preceded by an article, the relevant constraints rule out all the verbal readings of the word on this basis (see Table 1). Ideally, the input text should be disambiguated so that none of its constituent words have more than one morphological interpretation.

Table 1. Disambiguation of the word ‘lack’ with the help of Eng CG

“(“<the>”
 (“the” ART))
(“<lack>”
 (“lack” V SUBJUNCTIVE)
 (“lack” V IMP)
 (“lack” V INF)
 (“lack” V PRES -SG3)
 (“lack” N NOM SG))”

The word ‘lack’ can have both verbal (SUBJUNCTIVE, IMP, INF, PRES) and nominal (NOM) readings. However, if it occurs after an article (ART), all possible verbal readings can be eliminated. Consequently, one can apply to the ambiguous word ‘lack’ the constraint ((@w =0 (V) (-1C DETERMINER) that discards (=0) all lines containing the feature ‘V’ from an ambiguous cohort if the first word to the left (-1) is an unambiguous (indicated by the letter C) determiner (the feature ART being a member of the set DETERMINER) (Voutilainen, Heikkilä, and Anttila 1992: 2).

Once the parser has disambiguated the morphologically ambiguous words, the next task is to give them a *surface* syntactic analysis (Karlsson 1990).[7] The output of the morphological disambiguation module becomes, in turn, an input to the next module, namely, syntactic mapping; and this subsequently assigns all possible (surface) syntactic functions to each accepted morphological reading. Once again, a certain word form can have several different surface syntactic functions. A noun, for instance, can be a subject, object, indirect object, and so forth. Consequently, to give each word its correct syntactic reading, syntactic constraints are applied after mapping. These constraints are similar to morphological disambiguation constraints in that they discard contextually illegitimate syntactic function tags (see Voutilainen 1994: 16–17). The final output of the parser is a text in which, in a best-case scenario, each word is labeled (or 'tagged') with its correct morphological reading and syntactic function.

Layers of Representation

A CG parser for a certain language is a set of constraints written with the CG formalism for that language and implemented as a computer program. The algorithm is language-independent and has been used to write grammars for different languages. Because of the way the CG parser for a particular language is made, its construction can be approached as a continuous process of representation involving different representative layers and repeated testing with corpora. In order to create a (grammar-based) parser for a particular language, one has to describe the rules of that language with the aid of a parsing formalism, a sort of metalanguage (such as CG). This set of rules (which are actually constraints in the case of CG) is, in turn, implemented as a computer program, whose code constitutes another layer of representation. In practice, the set of rules and computer program have been developed concurrently with the help of previously annotated text corpora. The work of representation is time-consuming and piecemeal – the first full-blown version of the EngCG parser (Karlsson et al. 1995) incorporated more than 2,000 rules, and a group of linguists and computer scientists needed several years to develop it into an effective language-technological tool. The linguists developed different aspects of the parser, including word-class disambiguation, syntax, and lexicon, and the computer scientist programmed the EngCG parser with the C language.

Prior to building the parser, however, the problem of how to represent the rules of a language had to be solved. This involves developing a proper formalism, which is, in fact, the crucial epistemic challenge in making a parser. In an interview, Fred Karlsson, the original creator of the CG-formalism, said that he tried for many years to develop a syntactic parser by applying several different theoretically grounded linguistic formalisms before he realized that in order to make a functionally robust parser, he had to "turn everything upside down and try to do something by rejecting" instead of stating positively what might be possible. Thus, in contrast to traditional grammars, which are *licencing* grammars, a CG grammar is a *reductive* grammar. The idea behind it is contrary to that of formal grammars, such as Chomsky's Transformational-Generative Grammar. Instead of defining the rules for the formation of correct expressions in a language (L), the task becomes to specify constraints that discard as many improper alternatives as possible (Karlsson 1995: 10). Of course, one may ask whether the eventual success of this way of proceeding is crucially due to the fact that processing a list of commands is the sort of task typically undertaken by computers.[8] This serves to highlight how, in simulation, the theoretical and other properties of the modeled systems are adapted to the requirements of computation and to the affordances of the machines themselves.

To show to the scientific community that a grammar-based parser could really be designed along these lines, the Helsinki group set out to build a functionally robust parser for English. The CG grammar is written in a corpus-based fashion. One starts with a morphologically analyzed text for which one writes constraints that disambiguate the words of that text. After the resulting grammar is applied to the manually disambiguated benchmark corpus, the software of the system generates application statistics for each of the constraints. On the basis of these statistics and after identifying the mispredictions, the grammarian either corrects or removes old constraints or creates new ones. Then the cycle is repeated, making use of any new corpus evidence until, as one of the researchers put it, "the grammar approximates human performance sufficiently."

One critical phase in the work of representation underlying the EngCG-parser is the implementation of the grammar as a computer program, which poses a delicate epistemic challenge in itself.[9] The fate of the Finite State Parser, which is another parser the group was developing, illustrates the indeterminacy and ambiguity of the transformation that occurs during implementation. The Helsinki group had great expectations of the Finite State Parser, but these hopes never materialized. Despite the theoretical attractiveness of the finite state method in modeling language, the Finite State Parser proved to be technically impossible given the current computational capacity of machines. However, not one of the researchers I interviewed was certain whether this objective was an impossible task to begin with, or whether it might have been accomplished by other means. The researchers also discussed the properties of different programming languages in the implementation task. One researcher explained the merits of programming language C, with which the fastest implementation of the EngCG is made, in the following way:

> If you have to make a really tight code, then you have to use a language that is as near as possible to the commands of the processor … C is very close to the machine and yet it sustains the abstractions sufficiently. You have to choose a tool according to the task. Sometimes it is a shovel, sometimes it is tweezers. Here, because I am aiming at efficiency, I'm using C. It is old-fashioned, yes, and awkward. But on the other hand it is a … handicraft-language (Lauri Carlson, Interview).

From the point of view of representation, it is interesting to note that the linguist and the programmer actually represent entirely different things, even though they are building the selfsame artifact. This was pointed out by the computer scientist of the group:

> And then I have to recognize the [parsing] problem as one of those abstract problems, that is, what kind of computational world it relates to … In linguistic theory you try to make a description of the world and in computer science you try to erase it (Pasi Tapanainen, Interview).

Thus parser construction opens up a view of representation that differs greatly from the traditional one which focuses *on the relation between a ready-made model and its real target system.* According to the traditional view, a model is epistemically valuable if it, *in itself*, gives us an accurate depiction of its object. Here, on the other hand, although the parser is an outcome of the labor of representation, the parser itself is valued instead for what it produces. The accuracy of this output representation is more important for the researchers than the ability of the parsing process itself to realistically imitate human linguistic performance. The researchers I interviewed were all of the opinion that, above all, the parser must function well, which means that a parser must be able to carry out some of the same tasks (i.e., syntactic analysis) that humans can. To do this, parsers do not necessarily have to be 'psychologically realistic,' and it is highly probable that they will not be so.

I am speaking about the work of representation rather than about representation, because of the piecemeal character of the process of building the parser. In actuality, the process renders what is already described theoretically (i.e., word classes and syntactical functions) into another form (i.e., into constraints) with the help of a parsing formalism and subsequently programming language. It is not a question of representing the structure of language itself (as the traditional model of representation would have us believe), but rather of depicting in a specific way a certain kind of linguistic knowledge:

> In the end, that which is being modeled is nothing more than ... the sort of analyses a linguist – not a normal speaker but a linguist – would give for a certain data. ... It is modeled from a very close perspective or small angle. Maybe one could say that what we model is the way in which a linguist would analyze, on a certain level, texts of a certain language (Atro Voutilainen, Interview).

Consequently, the epistemic problem of parsing is not so much about finding out what something (i.e., language) is like, but *how something can be done* given certain resources (the computer providing the critical resource in this case). Thus, the problem concerns more the ways of representing themselves rather than what is represented. It directs our attention to the *process* of representation in which what we represent is already represented somehow. This insight leads us to question the representationalist supposition that it is the real target system *as such* that is being represented. It redirects our attention to the importance of various representative devices and methods as means of production instead of focusing exclusively on what the ready-made model – which, from this point of view, is already a complex representative achievement – is supposedly about. Important though this question is, the fact that it has received such exclusive attention so far has led to a certain mystification of the question of representation itself. One is left wondering what connects the representation and its supposed target system, and one loses sight of the process of representation that provides at least part of the answer being sought.

Theory and Expertise

If the EngCG-parser cannot be seen as a clear-cut representation of any part of the world that exists independently from our representational endeavors, its relationship to theory is also not straightforward. In addition to treating models as representations, the philosophical tradition often considers models as interpretations or some sort of concretizations of (formal) theories, but this captures poorly the part that theory plays in language technology. In fact, the CG parser as a model is not an application of any distinct autonomous theory. Due to the corpus-based way it is created, the CG grammar does not resemble the ordinary grammars of any given language, and the constraints are quite different from the grammar rules described by syntactic theories. Instead of one general rule in the CG grammar, "there will be a few dozens of down-to-earth constraints that state bits and pieces of the phenomenon, approaching or even reaching the correct description by ruling out what cannot be the case in this or that context" (Karlsson 1995: 26).

There are also certain parsers based on autonomous grammar theory, but their success in parsing running texts has been rather modest (see Black 1993). The problem has been that the numerous distinctions made in these grammatical representations of language require a certain semantic and higher-level knowledge of language that is not readily available (Voutilainen 1994: 30–31). Because modeling can be seen as a continuous process of representation, the problem here can probably be interpreted as one of transformation. Grammatical representations that are explicitly bound to a certain formal theory fail because such representations are not unambiguously transformable to an empirical level.

It should be noted that a parser such as EngCG should not be considered to be an atheoretical entity, despite not being a concretization of any autonomous formal theory. It nevertheless embodies linguistic concepts such as dependence, word class, and syntactical functions in its construction and output. In fact, in the field of language technological research, the Constraint Grammar is presented as a theory of how to design a doable parser – it is a description of the principles needed for inducing the grammatical structure of the sentences by automatic means. In the theoretical presentation of the Constraint Grammar (Karlsson et al. 1995), these principles are given in the form of a discussion of twenty-four parsing goals (e.g., G-10 "Morphological and lexical analysis is the basic step of parsing") and seven specific strategies (e.g., S-1 "Disambiguate as soon as possible") (Karlsson 1995).[10] The very generality of these principles, however, serves to underline the importance of a specific kind of expertise in parser construction.

Expertise has typically been attributed to local, experience-bound, and tacit dimensions of science, but it has remained rather difficult to tell what expertise consists of and how it is born (see, e.g., Collins and Evans 2002). The parser-building process described above suggests that expertise is tied importantly to certain specific methods and artifacts, and that it involves many kinds of knowledge and skills and an ability to synthetize them in view of certain goals. These goals should not be understood as pregiven or fixed. It also appears that an important part of expertise is the ability to define goals in such a way that they are doable. In this sense, expertise can be considered as object-bound knowledge and skills that emerge in given activities with their own characteristic tools and goals.

The epistemic challenge and the special kind of expertise needed in parsing were expressed very well by the response one interviewee gave to my question of whether building a parser is a practical problem (because linguistic theories do not play any noticeable part in it):

> Building a proper parser is not a practical problem at all. If it were, it would have been solved a long time ago. It's not a question of lacking effort or that not enough work was invested in it. ... It is in a certain way a genuinely difficult problem that is difficult to explain to anybody who does not know the structure of language. ... [Because of the complexity and ambiguity of language] the combinatorics explode. In Finite State parsing, we even had 10 to the power of 100 choices to choose from, so that we were infinitely far away from anything that could be computed, and because of that you have to have a certain strategy to attack this difficult problem (Kimmo Koskenniemi, Interview).

As this comment shows, the epistemic challenge of building a parser is to find a strategy for or an idea on how to take into account both the complexities of language and the computational capacity of computers in view of the task to be accomplished. This involves an ability to deal simultaneously with different kinds of knowledge and to find out what is doable given the resources at hand. One interviewee pointed out that, in his opinion, Karlsson's main achievement was exactly his insight in setting the goal of syntactic parsing to a doable level, to that of surface syntax, instead of aspiring to more theoretically bound, deeper levels. Several researchers also stressed the importance of methodological know-how; the ability of a researcher to (fore)see what can be done with different methods.

Even though the invention of a parsing formalism and the setting of an appropriate goal for parsing were crucial for the success of the EngCG parser, much must also be attributed to the specific competencies of the linguists and computer scientists who actually made it. The efficiency of parsing, which is one of the most important features of a parser, is largely due to implementation and thus to the programming skills of the computer scientist in question. But even more than that, the interviewees emphasized the respective skills of the linguists making the grammar. Once again, these skills consist importantly in the ability to use and combine different kinds of mostly linguistic knowledge about the structure of language and its use. Moreover, the expertise is coupled with the artifact and cannot be easily detached from it. Consider Voutilainen's reflections on the learning process that occurs during the creation of a grammar:

> In the making of a grammar it's not so much a question of how well you know the language you are examining, it's more or less like programming. ... It is as if you just saw the generalisations there. What I have learned, at least, is to think through that formal grammar ... through that descriptive machinery. In a way it is a question of getting into a kind of rut, too ... (Atro Voutilainen, Interview).

From the philosophical point of view, the language technologists also displayed a rather instrumental attitude toward parsing. For instance, even though the rule-based and probabilistic approaches to parsing seem to imply very different suppositions concerning language and cognition (i.e., language as a categorical vs. probabilistic phenomenon, and cognition as symbol manipulation vs. pattern-matching), the language technologists were typically more interested in the respective results and properties of different modeling architectures. To be sure, in the 1990s, there was an occasionally heated debate between the proponents of the two paradigms (see Voutilainen 1999b: 218–219). The outcome of that discussion has been the realization that both paradigms have their strong and weak points. For instance, building state-of-the-art statistical taggers needs carefully annotated training corpuses (preferably a few thousand words), but then the tagger can be trained with very little human effort, whereas writing a grammar is relatively time-consuming (requiring several months of rule-writing and testing). This seems to speak in favor of the statistical method. Yet, if the input text represents a domain or style not adequately represented in the training corpus, the statistical tagger's accuracy is likely to deteriorate, whereas correcting and improving the grammar of a linguistic tagger (e.g., for customizing the system for the analysis of texts from some particular domain) is not particularly difficult. Different kinds of hybrid approaches have been developed once "the most proponents of both camps have recognized that the 'enemy' is making useful contributions to the development of tagging" (Voutilainen 1999a: 20).

THE PARSER AS AN EPISTEMIC ARTIFACT

A parser is an example of a model whose scientific status is neither predominantly tied to its representative function nor based on any pre-existing autonomous formal theory. In assessing its epistemic value, it seems to me that one promising approach is to treat it as a constructed thing, an artifact, that can give us knowledge in various ways and that also, in itself, provides us a new object of knowledge. Elsewhere, I have argued that models in general can be treated as epistemic artifacts (Knuuttila and Voutilainen 2003; Knuuttila 2004).

Approaching models as epistemic artifacts draws attention to both their intentional and material dimensions from the interplay of which their epistemic qualities arise. Models gain their intentionality by being constructed, used, and interpreted in purposeful human activities. On the other hand, there is nothing to use, construct, or interpret unless models are materialized in some medium. This material dimension of models makes them able to mediate and travel between different groups, epistemic activities, and disciplines, and this, in turn, accounts for their *multiplexity* (Merz 1999). In fact, it is typical of modeling that the same computational templates travel across sciences (see Humphreys 2002, 2004). As they gain different interpretations and uses, they become part of the embodiment of different models.

The concept 'epistemic artifact' draws attention to the fact that in modeling, we are devising artificial things of which we know either the structure or the initial conditions and the mechanism. The traditional way to approach models would be to claim that these structures or mechanisms stand for something in the real world that exists independently of its representation. I call this an observational approach to the epistemic value of models. It keeps the artificial and the real separate from each other, and claims that the artificial gives us knowledge of the reality because it happens to depict the real things correctly in some respect. This approach abstracts away from the representative and interpretative work that goes on in science, losing sight of the cognitive point of modeling. Even if we aimed to model a certain external phenomenon in order to acquire knowledge of it – which need not be the purpose of our model, as the case of parsers shows – modeling proceeds rather by representing something tentatively than by isolating or abstracting some 'essential' features of the world as if we knew them already!

In treating models as investigative instruments, Morrison and Morgan (1999) break away from the observational approach to models and move toward a more interactive approach according to which the epistemic value of models can be seen to emerge in the processes of building and using them (instead of "using" models, Morrison and Morgan talk about "manipulating" them, probably because most of their examples are of theoretical models in physics and economics). Because of its focus on how we learn from models by building and manipulating them, this approach in fact implies that there needs to be something more concrete to work on than just conceptual ideas. For us to learn from a model, it has to be a self-contained thing with which we can interact.

Building on this approach, I prefer, however, to go one step further and treat models as artifacts, because I feel that this leads us to deal more concretely with the actual processes of modeling. Moreover, the word 'artifact' underlines the general claim of this paper that the epistemic value of models can be attributed to their productivity rather than to their being straightforward representations of something. It also seems to me that this artifactual and productive approach makes sense especially in the context of computer modeling. It makes allowance for the ineluctable materiality of the computer itself, the laborious work of writing the code, and the genuinely cross-disciplinary computational methods that are used to model widely different things, all of which are important characteristics of simulation modeling.

In which ways are models then productive? If we consider the EngCG-parser as an epistemic artifact, its scientific value can be approached through its different roles as a tool, object, and inferential device. As *tools*, parsers are valuable for both their product – a tagged text – and as a first stage in many natural language processing (NLP) systems. Tagged corpora are needed for many purposes such as corpus-based dictionaries[11] and grammars as well as corpus-based linguistic research. In the development of language technology, parsers are, in turn, typically used as modules in a variety of NLP applications, such as automatic term recognition, automatic thesauri, and speech technology appliances. Furthermore, not only linguists use parsing methods. Parsing formalisms might also be useful for the analysis of any symbol string, such as a DNA sequence.

One could, of course, ask whether these instrumental uses of parsers really make them valuable epistemic objects in their own right. I think they do. In discussing epistemic things, Hans-Jörg Rheinberger notes how any experimental system is defined by the technical tools it employs, through their impact on "a new generation of emerging epistemic things" (Rheinberger 1997: 30). This applies also to the role of parsers in the development of new NLP systems. Rheinberger notes further that the difference between a technical tool and an epistemic thing is functional rather than structural, "depending on the place or 'node' it occupies in the experimental context" (p. 30). Indeed, parsers function not only as tools but also as *objects* of language technological research.

The strong focus on the representative aspects of models makes it easy to forget that the models and computational methods used in science are interesting and important research objects in themselves. Much of our knowledge actually concerns the artifactual sphere with which we have surrounded ourselves, instead of natural systems. As soon as we start building parsers, for example, their fabrication becomes an interesting epistemic and methodological issue in its own right. We have, in fact, created new objects of research that spawn a host of new problems: The analysis of words in the lexicon of a parser is not an inconsequential problem, and neither is the problem of how to implement the parser so that it works more quickly and is more space-effective. Furthermore, one needs to find out what kind of information is needed for correct disambiguation, how this information can be acquired (e.g., on the basis of observations by linguists, automatic learning algorithms, or combinations thereof), and how such information should be represented (e.g., in the form of rules, collocational matrices, hidden Markov models, or neural networks).

The epistemic value of a parser as an *inferential device* derives largely from its being a constructed thing whose principles, mechanism, or structure we know. In this regard, parsers resemble theoretical models. The epistemic functioning of both computer and theoretical models is due to the constraints and affordances they embody. Interestingly, the constraints and the affordances of models are intertwined, and it is difficult to tell them apart. One might say that the constraints built into a model afford various kinds of inferences. In the case of parsers, the constraints are made operative by implementing the language description as a computer program, and this has actually made grammars more interesting as scientific objects. Traditionally, the evaluation of grammars has relied on conventional academic discussion, but it is practically impossible for humans to consistently follow such complex rule systems as extensive grammars. Thus the computer makes the evaluation of grammars easier by providing an interpreter with well-known knowledge sources and well-known operation (Knuuttila & Voutilainen 2003). Due to the implementation of the grammar, the linguist also has the possibility to learn from the performance of the model by finding out which rules cause trouble and by trying out different ones – which stresses the epistemic importance of the workability and 'experimentability' of models. Finally, it seems clear that making functional language-technology tools has taught us a lot about language, for instance, about the regularities of our language use and the polysemous and ambiguous nature of our linguistic 'order.' In that sense, language technology has had at least an indirect impact on more theoretically inclined linguistics.

One might want to categorize only the grammar underlying the parser as a model – which would be in line with the semantic conception of models as abstract structures – and then ask whether these constraints taken as a model of language correctly represent our linguistic competence or the structure of language. But because of the computational complexity of the task, this and many other questions cannot be answered unless the set of constraints is implemented as a computer program. Moreover, the formal language with which the constraints are described is devised with its implementation – and workability – in mind. Consequently, rather than being a linguistic description that is interesting as such, the set of constraints underlying the parser should instead be considered as a part of the parser.

Focusing on the parser rather than only on the set of constraints underlines the importance of the instrumental fitness of the parser for creating new epistemic links. Once the parser functions well, it provides an interesting starting point for diverse interpretations and questions. One can study, for instance, what properties distinguish successful models from unsuccessful ones, what assumptions can be made about human language faculties, and so on. Indeed, one of the criteria for assessing parsers is ascertaining how 'realistic' they are, even though they are not considered to be complete representations of human linguistic competence. Moreover, the fact that a working parser can be made by eliminating rather than licensing supports the traditional (pre-Chomskyan) view that a language is an open-ended system with no strict demarcation lines between what is grammatical and what is not. These examples show that most of the evidence and insights into language and cognition with which parsers provide us are indirect – linking, as they do, to other bodies of knowledge.

Consequently, a parser, much like many other models used in science, seems to function more like an inferential device than as a straightforward representative entity.

CONCLUSIONS

I have argued that viewing models as productive entities gives us a more full-fledged picture of their epistemic value than the traditional focus on their representative function. As an example, I have used the Constraint Grammar parser, which, at first sight, seems to be relatively uninteresting from the epistemic point of view. However, approaching the Constraint Grammar parser as an epistemic artifact actually demonstrates how many roles it has in science and how it already in itself presents us with a new kind of reality. Models need not be treated just as surrogates for reality proper.

In stressing the epistemic value of the materiality, and the consequent workability and 'experimentability' of models, the artifactual and productive approach to models set forth in this paper differs significantly from the traditional philosophical accounts of models. These accounts have more often than not been predisposed to treat models as abstract, idealized, or theoretical structures that *stand for* real target phenomena. The tendency to treat representation as *the* epistemic task of models derives from this point of view. If models were but abstract structures, it would be difficult to understand how they could give us knowledge except by representing the world more or less accurately. But, on the other hand, if models are recognized as material and productive things, then it is evident that they already provide us with something tangible to work on and experiment with. This, in turn, speaks for the indispensable and versatile role of artifacts as regards our cognitive endeavor: Because the nature of our language is not intuitively known to us, the only way to gain knowledge here (as well as anywhere else) is mediated through the creation and use of artifacts.

The main thrust of this article is not, however, directed against representation per se, but rather at a certain conception of representation and its place in scientific endeavor. I have suggested that a close look at the practices of modeling provides a new angle from which to approach representation. Rather than being representations in themselves, models are often valued for the results they produce. Yet they are entities created by representing some initial conditions, mechanisms, rules, or structures. Both of these properties of models are especially salient in the case of simulation models. On the one hand, simulation models are an outcome of complicated representative procedures. These procedures are not reducible to any underlying theory, but make use of knowledge and expertise of different kinds. On the other hand, simulations are characteristically valued for their output representations.[12] Thus, being productive things created by representation, simulation models question the distinction between the performative and representational conceptions of science and challenge us to approach representation performatively. From this point of view, representation is less a relation to be aspired to by philosophical analysis than an important object of factual knowledge itself.

ACKNOWLEDGMENTS

I wish to thank all my interviewees, of whom Lauri Carlson, Fred Karlsson, Kimmo Koskenniemi, Pasi Tapanainen, and Atro Voutilainen have been quoted in this article.

University of Helsinki, Finland

NOTES

[1] The simulations and simulation models I am talking about in the following are *computer* simulations.

[2] See Küppers, Lenhard, and Shinn in this volume.

[3] The empirical part of this study is based on 24 semistructured, 1.5- to 3-hour long transcribed interviews that I conducted during the years 2000–2004. I interviewed a total of 16 researchers and representatives of language technological companies who have either been doing long-term language technological research in the Department of General Linguistics at the University of Helsinki or have otherwise been affiliated with the group. The written material on which the study is based consists of publications by the researchers interviewed, and of reports and documents concerning the Research Unit for Multilingual Language Technology at the University of Helsinki.

[4] On the different approaches to simulation, see Küppers, Lenhard, and Shinn in this volume.

[5] These reminiscences are of course not accidental: Chomsky emphasizes the importance of the "technical devices for expressing a system of recursive processes" for the development of his grammatical theory (1965: 8).

[6] Depending on the language, the disambiguation problem differs widely. In English, it is indeed typical that many familiar words can be both nouns and verbs, and, in many cases, they can have other readings as well. It has been estimated that nearly every second word in English is thus *categorially* ambiguous (DeRose 1988).

[7] Surface syntactic functions deal with the actual written words and do not attempt to reach potentially deeper, theoretically motivated levels.

[8] I owe this important insight to Johannes Lenhard.

[9] For the implementation of EngCG Grammar, see Tapanainen (1996).

[10] Thus the 'theory' here can be considered as a set of general instructions, resembling somewhat the general-equilibrium approach in macroeconomics described by Marcel Boumans in this volume.

[11] For instance, the entire *Bank of English*, a large text bank containing 200 million words launched by COBUILD (a division of HarperCollins Publishers) and The University of Birmingham, has been tagged and parsed in Helsinki using the TWOL and CG computer programs (see Järvinen 1994).

[12] This discussion has mostly concerned the different merits of visualization (see, e.g., Hughes 1999; Humphreys 2004; Winsberg 2003).

REFERENCES

Black, E. (1993). "Statistically-based computer analysis of English", in E. Black, R. Garside, and G. Leech (eds.), *Statistically-Driven Computer Grammars of English: The IBM/Lancaster Approach*, Amsterdam: Rodopi, pp. 1–16.

Chomsky, N. (1957). *Syntactic Structures*, The Hague, NL: Mouton.

Chomsky, N. (1965). *Aspects of the Theory of Syntax*, Cambridge, MA: MIT Press.

Collins, H.M. and R. Evans (2002). "The third wave of science studies: Studies of expertise and experience", *Social Studies of Science*, **32**(2): 235–296.

DeRose, S.J. (1988). "Grammatical category disambiguation by statistical optimization", *Computational Linguistics*, **14:** 31–39.

French, S. (2003). "A model-theoretic account of representation (or, I don't know much about art…but I know it involves isomorphism", *Philosophy of Science*, **70**: 1472–1483.

Frigg, R. (2003). *Re-presenting Scientific Representation,* PhD thesis, London School of Economics, September 2003.
Galison, P. (1996). "Computer simulations and the trading zone", in P. Galison and D.J. Stump (eds.), *The Disunity of Science*, Stanford, CA: Stanford University Press, pp. 118–157.
Giere, R.N. (2004). "How models are used to represent physical reality", *Philosophy of Science*, **71**: S742–S752.
Hacking, I. (1983). *Representing and Intervening*, Cambridge, UK: Cambridge University Press.
Hughes, R.I.G. (1997). "Models and representation", *Philosophy of Science*, **64:** S325–S336.
Hughes, R.I.G. (1999). "The Ising model, computer simulation, and universal physics", in M. Morgan and M. Morrison (eds.), *Models as Mediators*, Cambridge, UK: Cambridge University Press, 97–145.
Humphreys, P. (2002). "Computational models", *Philosophy of Science*, **69**: S1–S11.
Humphreys, P. (2004). *Extending Ourselves. Computational Science, Empiricism, And Scientific Method*, Oxford, UK: Oxford University Press.
Järvinen, T. (1994). "Annotating 200 million words: The Bank of English Project", in Proceedings of COLING-94, vol. 1 (Kyoto), pp. 565-568.
Karlsson, F. (1990). "Constraint grammar as a framework for parsing running text", in H. Karlgren (ed.), COLING-90: Papers Presented to the 13th International Conference on Computational Linguistics, Helsinki, pp. 168–173.
Karlsson, F. (1995). "Designing a parser for unrestricted text", in F. Karlsson, A. Voutilainen, J. Heikkilä, and A. Anttila (eds.), *Constraint Grammar. A Language-Independent System for Parsing Unrestricted Text*, Berlin: Mouton de Gruyter, pp. 1–40.
Karlsson, F., A. Voutilainen, J. Heikkilä, and A. Anttila (eds.), (1995). *Constraint Grammar. A Language-Independent System for Parsing Unrestricted Text*, Berlin: Mouton de Gruyter.
Knuuttila, T. (2004). "Models, representation and mediation", a paper presented at PSA 2002, http://philsci-archive.pitt.edu/. To be published in *Philosophy of Science*.
Knuuttila, T. and A. Voutilainen (2003). "A parser as an epistemic artifact: A material view on models", *Philosophy of Science*, **70**: S1484–S1495.
Koskenniemi, K. (1983). *Two-Level Morphology. A General Computational Model for Word-form Production and Generation*, Publications of the Department of General Linguistics, University of Helsinki, No. 11, Helsinki: University of Helsinki.
Merz, M. (1999). "Multiplex and unfolding: Computer simulation in particle physics", *Science in Context*, **12**: 293–316.
Morrison, M. and M.S. Morgan (1999). "Models as mediating instruments", in M. Morgan and M. Morrison (eds.), *Models as Mediators*, Cambridge, UK: Cambridge University Press, pp. 10–37.
Pickering, A. (1995). *The Mangle of Practice: Time, Agency and Science*. Chicago, IL: University of Chicago Press.
Rheinberger, H.-J. (1997). *Toward a History of Epistemic Things*, Stanford, CA: Stanford University Press.
Tapanainen, P. (1996). *The Constraint Grammar Parser CG-2*, Department of General Linguistics, Publication No. 27, Helsinki: University of Helsinki.
Teller P. (2001). "Twilight of the perfect model model", *Erkenntnis*, **55:** 393–415.
Voutilainen, A. (1994). *Three Studies of Grammar-Based Surface Parsing of Unrestricted English Text*, Department of General Linguistics, Publication No. 24, Helsinki: University of Helsinki.
Voutilainen, A. (1995)."Morphological disambiguation", in F. Karlsson, A. Voutilainen, J. Heikkilä, and A. Anttila (eds.), *Constraint Grammar. A Language-Independent System for Parsing Unrestricted* Text, Berlin: Mouton de Gruyter, pp. 165–284.
Voutilainen, A. (1999a). "A short history of tagging", in H. van Halteren (ed.), *Syntactic Wordclass Tagging*, Dordrecht, NL: Kluwer Academic Publishers, pp. 9–21.
Voutilainen, A. (1999b). "Hand-crafted rules", in H. van Halteren (ed.), *Syntactic Wordclass Tagging*, Dordrecht, NL: Kluwer Academic Publishers, pp. 217–246.
Voutilainen, A., J. Heikkilä, and A. Anttila (1992), *Constraint Grammar of English: A Performance-Oriented Introduction,* Department of General Linguistics, Publication No. 21, Helsinki: University of Helsinki.
Winsberg, E. (1999). "Sanctioning models: The epistemology of simulation", *Science in Context*, **12**: 275–292.
Winsberg, E. (2003). "Simulated experiments: Methodology for a virtual world", *Philosophy of Science*, **70**: 105–125.

CHAPTER 4

MICHAEL HAUHS* AND HOLGER LANGE**

FOUNDATIONS FOR THE SIMULATION OF ECOSYSTEMS

INTRODUCTION

Interactive simulation of ecosystems is a new computational technique that extends the scope of information technology (IT) applications into ecology. As with every newly introduced technology, it has the potential of changing the problem perception within its field of application. What appears now as a technically solvable problem, and what remains an unsolvable problem – technically or in principle? By their successes and failures, simulation models may change the attitudes toward ecosystems not only in science but also in ecosystem management. The relationship between ecosystem practice and research is usually a problematic one (Peters 1991; Beven 2001; Bocking 2004; Kimmins et al. 2005). Successful utilization schemes predate ecology and ecosystem research, as can be seen, for example, in the "plenterforests" of Central Europe (Schütz 2001).

Ecosystems commonly fall under the rubric of complex systems (West and Brown 2004). Nevertheless, in the practical management of certain ecosystems, we encounter simple heuristic rules of human interference that are often derived from cultural traditions rather than from scientific study. The increased technical power of computer-based simulation tools and their increased mathematical formalization may either remove former technical limits (e.g., of prediction) or, in contrast, reveal the fundamental character of some of these limits. Here, we shall argue that both cases occur, and that the main effect of simulation technology is to bring the distinction between these cases into scientific awareness.

This chapter is organized as follows: First, we clarify our terminology to demonstrate that we are actually introducing a new modeling paradigm, and exemplify its domain of application. Then, we briefly review the traditional algorithmic modeling paradigm for state-based systems before discussing interactive simulation as an extension to this based on the mathematical notion of streams as an abstraction of behavior. We try to show that interactive simulation becomes especially useful when applied to models of living systems and ecosystems. Finally, we discuss the different limits encountered in genuine interactive behavior and displayed by genuine complex systems. Although both notions can be used to address theoretical and practical limits

J. Lenhard, G. Küppers, and T. Shinn (eds.), Simulation: Pragmatic Construction of Reality, 57–77.

of simulating ecosystems, currently, complexity is used exclusively for demarcating limits of simulation models (Ulanowicz 2004). Interactive simulations may explain the empirical simplicity that is often encountered in ecosystem management. Thus, they ought to play an enhanced or even dominating role in theoretical as well as applied ecosystem research.

INTRODUCING TERMINOLOGY

For systems as well as models, we distinguish between algorithmic and interactive types of *behavior*. Behavior addresses all kinds of temporal changes, both in an active and a passive sense. In particular, we use a notion of behavior that goes beyond dynamic systems theory and allows for active *choice-making* behavior within the systems considered. In dynamic systems theory, behavior is reduced to (algorithmic) functions of *state transitions* alone. The concept of state is a very prominent modeling abstraction developed in physics.[1] However, active choice-making behavior is impossible to incorporate in algorithmic models, although it is often encountered in living systems.

The Traditional Algorithmic Modeling Paradigm

In algorithmic models, functional behavior is reduced to structure, that is, the configuration of objects in (state) space and their change over time under the entailment of 'Natural Law' (Rosen 1991). The observed behavior in, for example, experiments can be *explained* or predicted algorithmically by a system of equations subject to specific boundary conditions. When one views the world from this approach, behavior is inevitably reduced to a secondary role, referring to state transitions governed by dynamics. Many examples of this approach and the relationships between structure and function are given in the introduction to this book.

In the cases of ecosystems and social systems, the observed structure appears to be irreducibly complex. Thus, any simplicity or regular behavior encountered in these systems appears surprising to 'reductionist science' and is typically lost in scientifically rigorous approaches to such systems. Ecosystem researchers confronted with some seemingly simple rules of ecosystem managers tend to ascribe their success to a system simplification obtained by taming; the main effect of, for example, agroforest monocultures is a reduced number of degrees of freedom (Bocking 2004). On the other hand, the predictive ability of models based on scientific process understanding (e.g., for a forest under climate change) is very low. This mismatch is a hint that the wrong modeling paradigm might be being used.

Where these systems are studied and simulated today in ecosystem research, the ultimate goal is to replace the heuristics of management with a process-based understanding of the dynamics (Lansing et al. 1998). Here, scientific knowledge in considered to be, in principle, superior to any other form of knowledge. The leading example in environmental and ecological sciences is meteorology, in which empirical models have been replaced successfully by physical models in operational weather prediction. The computational tools used for weather prediction express the current technical limit of the state model type introduced by Newton. Its solution was already recognized as a technical problem a century ago (Bjerkenes 1904). Attempts to sketch a similar path for the simulation of ecosystems (including the complete biosphere) appear much less convincing (Schellnhuber and Wenzel 1998), because they lack thorough real-world case studies. The application to climate models already has to deviate from a mechanistic recipe of dynamic state theory (see Lenhard's argument, this volume, and the introduction).

The Interactive Modeling Paradigm

Systems with relatively simple external behavior at a user interface and complex internal structure occur today in high-end human technology. A computer is only one out of many examples for designed systems that, while becoming increasingly difficult to built, are simpler and more robust to use. A computer is deliberately designed to provide a simple intuitive service. Recently, a number of theoretical approaches have been suggested to express formally 'what computer scientists do' when they build interactive or concurrent programs. These approaches are based on the notion of streams and are, mathematically speaking, algebraic duals to the traditional, algorithmic ones in computer science (Gumm 2003; Arbab 2005). Instead of seeking the model that provides the simplest explanation of a phenomenon by identifying an initial state, they search for the most comprehensive model of behavior in terms of sets of streams.

Examples for interactive behavior include cases in which the simulation is not derived from a comprehensive scientific understanding or reconstruction of the modeled ecosystem, but documents and communicates heuristic knowledge about (managed) ecosystems and how they have been actively sustained by human interference.[2] We shall argue below that such simulations have been established in other areas and may, in a more long-term perspective, change the foundations of ecological modeling.

Applications of interactive simulation models occur in many fields today. Prominent and well-established examples are chess computers or flight simulators. We shall argue that silviculture in forestry may provide examples of interactive simulation as well, and that this model type may be regarded as a fundamental one in terms of ecosystem research. We conjecture that interactive simulation models are qualitatively different from the model classes used in physics.

This chapter takes an 'engineering perspective' on interactive behavior: The (supposed) simplicity of ecosystem responses as perceived in traditions of, for example, hunting, farming, or silviculture provides us with a unique modeling challenge. There may be considerable human expertise (skill) present in any of these traditions, but skill[3] when evaluating and deciding on proper management is difficult to explain within a scientific context. Skilled behavior toward ecosystems can be referred to as 'tacit knowledge.' In traditional indigenous utilization schemes, it is part of an embedded relation with respect to the environment. It may appear in sharp contrast to scientific attitudes toward and perception of the same environment (Ingold 2000).

Where traditional systems of ecosystem management and land use have been studied in anthropology, the leading paradigm of the natural sciences has been criticized, and an extension to it has been proposed (Ingold 2000). We shall show below that the technical and theoretical extensions provided by computer science to the issues of interactive computation in the form of modern simulation models correspond to the concepts derived in anthropology to classify human culture: They all seem to aim at the (irreducible?) interactive aspects of these systems. It can be conjectured that earlier attempts by 'western' scientists to substitute indigenous forms of land use knowledge by dynamical models might have failed for principal reasons when truly interactive situations were involved. How much interactivity is implicated in a given ecosystem management scheme varies a lot and needs empirical testing. With the modern extension to IT, however, these situations can now be studied much more explicitly and locally. Silviculture in forestry serves as an example here.

When searching for the most appropriate simulation model class (algorithmic or interactive), we answer three parallel questions in the context of modeling the *behavior* of ecosystems:

- What is it that humans do, when they manage an ecosystem to fulfil a function[4] (and make a living)?
- What is it that scientists do when they study and model an ecosystem (to understand it, document and narrate its past; or estimate its future, evaluate its potential)?
- What is it that a computer provides, when modelers try to represent knowledge (managerial and scientific) and simulate an ecosystem?

We start with the definition of interaction proposed in computer science that provides us with a precise and sufficiently general notion of (choice) behavior in machines. The functional (algorithmic) behavior of machines will subsequently be recovered by imposing restrictions (i.e., algorithmic models are a special [limiting] case of interactive ones). An interactive model contains two elements:

- *streams* (Gumm 2003) serving as the mathematical representation of behavior, for example, of choice events that characterize an ongoing or already realized interaction, and
- a *real-world context* in which the outcome of any choice depends upon the sequence of choices made before. The outcomes of realized choices and the choices still to be made in the future are related through valuation and norms – technically, these introduce equivalence classes in the set of possible choices.

Typically, normative constraints will apply to choices. Take the following three examples:

- In chess playing, the options of winning should not be decreased as a consequence of the current choice.
- In airplanes, the options for a safe landing should not be decreased by an actual maneuver.
- In sustainable forestry, the options for further production and productivity should not be decreased by an actual thinning or harvesting decision.

In all three cases, the normative element stems from a predefined goal function (to win, to land safely, to sustain timber production). However, in computer science, a typical situation even lacks goal orientation, as in a persistent client/server interaction in which perennial service provision on the server side is mandatory but not an element of the 'goal' of the interaction with the client. It is obvious that the 'ultimate' outcome (after any finite time) of an ordered 'stream' of choices is fundamentally unpredictable for an algorithmic model. By definition, a precisely predictable situation[5] is noninteractive, and could only be considered as such by an observer ignorant of this algorithmic possibility. In any other situation, the choices taken become interspersed with evaluations of their outcomes and a reassessment of the altered potential for follow-up choices. This is one of the most important aspects by which the algorithmic and the interactive simulation models differ. In an algorithmic simulation, the predictive task is the challenging one, whereas evaluation of its results is relatively easy (again, weather prediction provides an illustrative example). In an interactive simulation, the evaluation task is the challenging one, whereas the (immediate) prediction of the results of the next choice is trivial, and the long-term prediction either appears to be or is impossible. Here, chess playing, silviculture, and pilot training share an interactive character demonstrating the need for a new abstraction in simulation models.

The reason for this difference lies in the relationship to the environment and not in the system itself. The options provided by the environment through an interactive interface may change as a result of past choices, and an observer *enclosed* by interactive interfaces meets an algorithmically unsolvable problem. For example, in plantation forestry, more trees are planted than will ever reach mature age. However, at the time of planting, which trees are to be harvested or taken out is left open for later thinning decisions. Here, the reason is the phenotypic plasticity of trees, or the genotype/phenotype distinction in general. Uncertainty with respect to the actual soil conditions (may vary due to spatial heterogeneity) or with respect to the actual weather conditions over a rotation period renders interactive decisions by foresters in many cases inevitable.

In flying, the interaction between different airplanes or with traffic control leads to another example in which choice situations posed by the environment are unpredictable but can be handled more efficiently through interaction. These choices require the proper training of human pilots in (interactive) flight simulators to reliably provide the necessary competence.

GENERALIZING NOTIONS OF BEHAVIOR: WHAT IS INTERACTION IN COMPUTER SCIENCE?

Models for computation are usually based on the notion of the Universal Turing Machine (TM). Recently, an extension of the TM to a *Persistent* Turing Machine (PTM) has been proposed by Goldin et al. (2004) and Wegner and Goldin (1999). Persistence is introduced by a read/write tape that is *not* reset to an initial state between subsequent computational cycles. The PTM interacts with its environment in the sense that later input from the environment may depend upon former output from completed computations of the machine. PTMs may imply a new and more general meaning for 'computing' than the TM. If we want to restrict the notion of computing to what is formalized by TMs, than it becomes unclear what extra services today's computers are able to provide (Arbab 2005).

This relationship between external interactive behavior and internal persistent memory states also holds outside computer science. It links memory and a privileged perspective from the inside with interaction on the outside. We apply it here to explain the pattern encountered in the success and failure of ecosystem modeling and simulation. What new capacities and limits of *interactive* computing can be expected to be relevant for describing and abstracting those real-world systems that have resisted progress in algorithmic computing so far? Whereas in the algorithmic model, one needs to distinguish computable from noncomputable functions, the limits of interactive models exist between unbounded and bounded forms of interaction depending on the question whether entirely new features and choices may appear at any time. Candidates for unbounded interactions are open-ended evolution (life) and open-ended communication (culture). In the realm of interactions, the cases in which the class of all possible choices has a finite (infinite) representation are termed bounded (unbounded). We see little or no progress in the capacity to predict ecosystems as computers become faster and able to handle observations from more complex systems (see Figure 1, upper part).

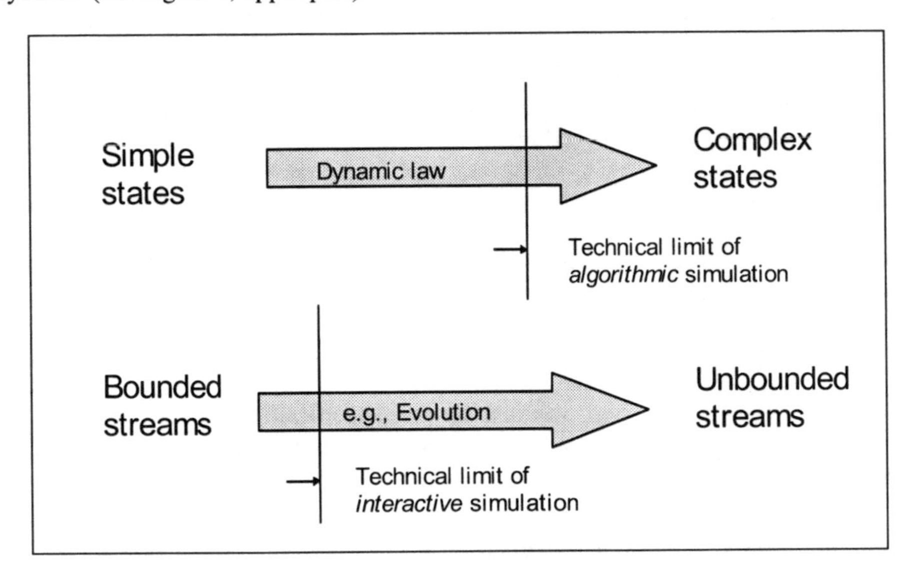

Figure 1. For algorithmic simulation models, a state may be too complex to be represented by a computable function (as e.g., in chaotic systems). For interactive models, the set of choices that produce entries in a (data) stream may be bounded or unbounded (as, e.g., in biological or cultural evolution). Only bounded sets can be represented in any model of an interaction

However, we may still expect progress when, for example, the relationships between forest growth models and a forester become represented by interactive models, and formerly unbounded situations can now be evaluated as bounded ones. All three examples above match this situation, whereas in chess, a winning strategy has remained algorithmically unattainable, the online access to the documented history of played out matches allows interactive simulation models to treat the choice problem as (almost) bounded (see Figure 1, lower part). In pilot training, flight simulators become better; the critical situations that may occur unexpectedly have become documented and are covered by the simulator. In silviculture, only the very initial steps in this direction have been taken, but, here as well, a bounded choice situation based on the historically documented and approved examples has become technically possible (Hauhs et al. 2003).

GENERALISING TERMINOLOGY: MODELING, COMPUTATION AND SIMULATION

We shall use the terms modeling, computation, and simulation in the following sense. *Modeling* is the most general activity, referring to a symbolic or virtual aspect of an investigated system in relation to its observed structure and/or memorized behavior.[6] Modeling is based on (consists of) mappings, termed 'representation' and 'implementation,' that establish relationships between a real, concrete realm of the world we live in and an abstract or virtual world providing the (partial) referents for models (see Figures 2 and 3). Social systems individually and collectively have access to the real world through observation and memory. They have established procedures (partly outside science) on how agreement can be achieved between different

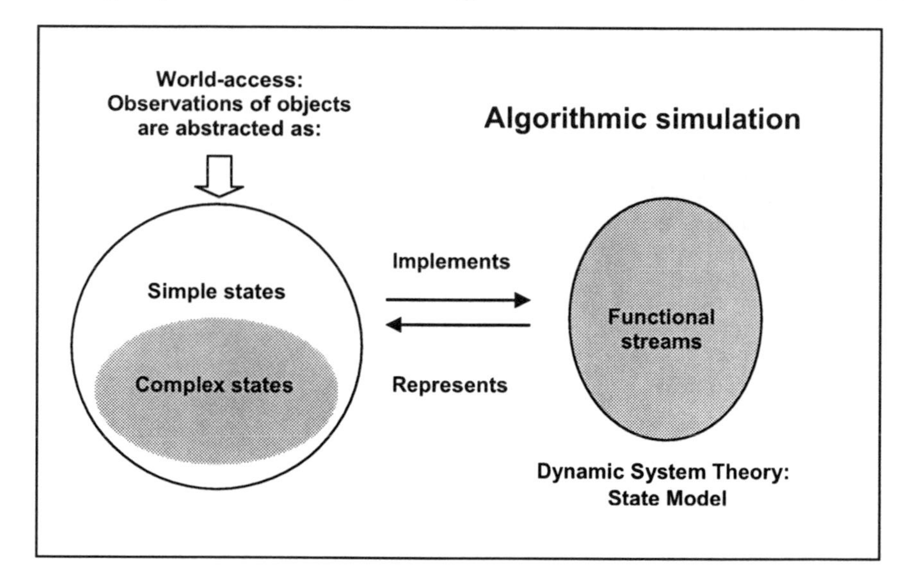

Figure 2. Relationships in scientific modeling (under the state model paradigm). In the real world, every system has a set of observables attached, which are represented as abstract state variables. Time variations in states are conceptualized as abstract functional *behaviors resulting from transitions under a dynamic law. Experiments in the real world can be conceptualized as local implementations of the dynamic laws. This traditional modeling paradigm will be referred to in the text as 'algorithmic computation.' A model in which the functional behavior is inferred from a 'faithful' representation of the observed states is termed an (algorithmic) 'computation.' A model in which the functional behavior is inferred without restrictions about the states is termed here an (algorithmic) 'simulation'*

individuals and subgroups over the content of their observations and memories. In a scientific context, access by (objective) observation is regarded as superior to access by memories (Rubin 1995; see Figure 2).

Modeling as a science needs to be open to testing, criticism, and revision. Scientific models are distinguished from nonscientific models by this grounding procedure in empirical knowledge agreed upon among a group of experts and open to critique. If, in addition, modeling can be a) formalized by mathematical structures or b) transferred into a (representational/symbolic) form in which steps are executed automatically by a computer, we will term this *computation.*[7] The sciences in which modeling can make use of an established (mathematical, systematic) theory are those that have developed *computational* branches, such as computational physics, computational chemistry, computational biology, or computational meteorology (the latter is not a standard technical term; we refer to computer-based weather forecasting as done routinely nowadays).

Other sciences, mostly those lacking underlying fundamental mathematical theories, use the label 'modeling' instead, as in, for example, ecological or environmental sciences. Besides scientifically based understanding, other forms of knowledge exist and are referred to as heuristics, skills, or tacit knowledge. Such forms of knowledge abound in ecosystem utilization and their respective management traditions (such as in hunting, agriculture, forestry, fisheries, gardening, etc.).

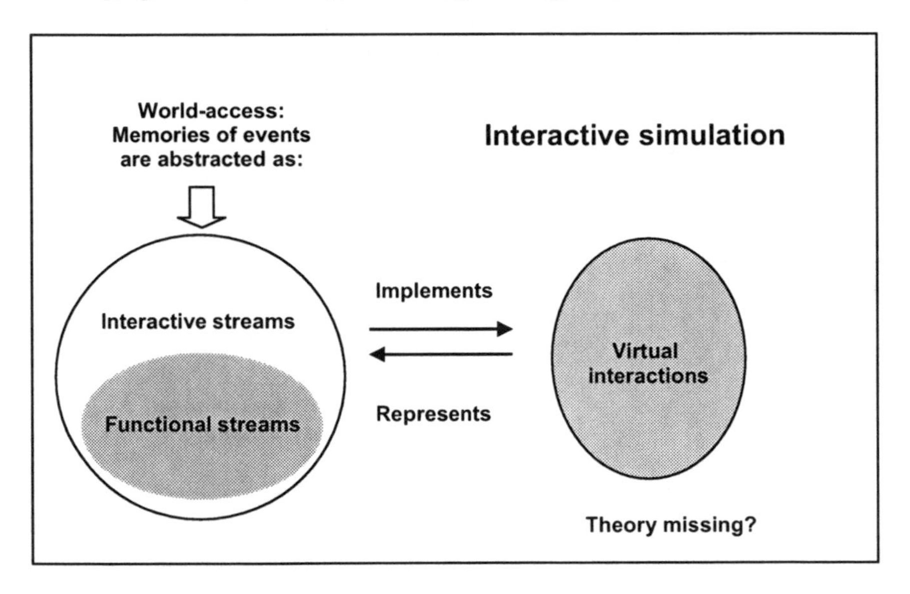

Figure 3. Relationships under the proposed second model paradigm. In the real-world system, time-ordered data streams are accessible by memory of agents. If the streams are generated interactively, they become represented as virtual choices in an interactive simulation (technically by persistent states of the interactive machine). Experienced time is conceptualized as choices realized by interacting partners subject to internal motives and social norms. This new modeling paradigm will be referred to in the text under 'interactive model.' It corresponds to a second and new notion of simulation, here referred to as interactive simulation

When the objective of the modeling is not to provide a realistic representation of the system's structure, but still to reproduce the observed behavior of target variables, the system is said to be simulated. In this usage, simulation means something different and, in terms of scientific rigor, *less* than computation. There is, however, a second use of the term *simulation*, and both of these usages have historic examples (see Terry Shinn's chapter in this book). The second case applies to situations related to interactive choices in which a body of knowledge changes only slowly relative to the lifetime of an expert. Relevant situations in which such tacit knowledge becomes relevant are sparse, and thus training to expert levels is a difficult task. Historical examples are generals' skills in battle tactics simulated in sandbox scenarios, or air force pilots' and gunners' skills in World War II simulated in analogue models of airplanes. Competence in chess playing simulated by modern computers can serve as another example or, as we shall argue below, many examples in ecosystem utilization such as silvicultural skills in forestry.

In this second sense, an (interactive) simulation approach represents *more* than a computational model (Figure 3). Despite the fact that the examples above appear to be modeled less rigorously when viewed in terms of the traditional approach, interactive simulations are practically without any rivals. In fact, they have added something new to the respective application field that is not yet properly accounted for in the foundations of modeling based on dynamic systems theory.

DEFINITION OF LIFE AND ECOSYSTEMS

Life is a phenomenon occurring at scales between macromolecules and the biosphere. Neither the molecular building blocks of a cell nor the global cycles of life's resources (e.g., of water) are alive. Physical processes can be used to delineate living systems from larger spatial scales by (noninteractive) functional behavior downward and from its simple (noninteractive) building units upward.

First. we shall try to define life and ecosystems exclusively using the terminology of the algorithmic model (dynamic systems theory). This represents an approach to define and relate terms by using established physical notions (i.e., reductionism). In this perspective, life is regarded as a phenomenon requiring a minimal complexity in order to execute or perform typical behavioral features such as self-reproduction, the abilities to adapt, evolve, and so forth. In this context, the potentially interactive character of these behaviors is (implicitly) abstracted away, or simply placed into the eyes of the observer. Above the complexity threshold and when provided with the appropriate conditions, the emergence of life may then become inevitable. Molecular biology seeks to identify minimal forms of living entities, whereas systems biology is often viewed as an attempt to compute or synthesize according to this modeling paradigm. These attempts have not been successful yet; in other words, the first living organism synthesized in the lab from molecular building blocks has still to arrive ("we are missing something fundamental" [Brooks 2001]).

At the other end of the scale, ecosystems are open to their environment and contain life. They do not live themselves but consist of living entities and abiotic constituents. *Ecosystem* is probably the most popular term among ecologists. It is defined only vaguely and carries many different meanings (even if we restrict its use to ecology alone). The openness toward its external environment can be described in (noninteractive) physical terms, for the living aspects we shall use interactive behavior as described in ecology. Thus, external boundaries delineating an ecosystem will be based on physical aspects, whereas delineation of behavioral epochs will be based on biological aspects.

Here we regard minimal ecosystems as the smallest evolvable living units that exchange abiotic fluxes of matter and energy with the environment in a noninteractive manner. When living systems become aggregated in the form of whole landscape units (such as watersheds), their behavior at the boundaries often becomes relatively simple and can be simulated algorithmically; matter and energy fluxes across watershed boundaries are functional (i.e., noninteractive). External relationships of such functional units can be studied only on the basis of the *physical* concept of interaction (*Wechselwirkung* in German).

In our terminology, an ecosystem is a noninteractive unit of a landscape (Pittroff and Pedersen 2005). The observed behavior of such units, however, appears as an anomaly in terms of hydrological transport models. It has not been possible to explain (uniquely reconstruct) runoff data by physical models (i.e., distributed hydrological models). Algorithmic simulation models are typically overparameterized with respect to the observed runoff data. The 'true' internal transport mechanisms needed to perform typical transport models cannot be identified directly from data. Within the algorithmic modeling approach, these difficulties are discussed in terms of the broad heterogeneity of hydrological catchments and the technical limitations of proper sampling.

To summarize, at the lower and upper cutoff scale of life, the modeling approaches based on algorithmic (simulation) models have, to our knowledge, not yet led to a 'living reconstruction' or a nontrivial prediction. There is a widely accepted explanation of these difficulties: Living entities and ecosystems appear to be (too) complex. However, empirical modeling of a runoff signal hardly involves more than two or three parameters (Jakeman and Hornberger 1993). This appears as an anomaly for algorithmic simulation models. Why does the complex system provide us with simple responses that turn out to be of particular interest for human utilization? We shall turn to the second modeling paradigm (interactive models) to seek more consistent answers to these questions.

We have suggested considering life as an irreducible interactive phenomenon. In order to extend this proposal to the ecosystem scale, we have to generalize the primitives used in the above ecosystem definition: Fluxes across the boundary will be generalized to streams.

STREAMS AND FLUXES

The basic mathematical notion used here for describing the boundaries of an ecosystem with an abiotic environment is a (data) *stream*.[8] A stream is a potentially infinite ordered (time) series of instances of discrete events – in our case, we are interested in abiotic events at the boundary of an ecosystem. With this application in mind, streams consist of transport events of extensive variables (matter, energy). Our primary example is water transport. Hence, we are seeking boundaries that are related to precipitation and runoff. These data streams can be described in a continuous or a discrete way; the latter being the canonical choice for digital computers. Also, measurements that have to respect the finite sensitivity of the instruments are inevitably discrete. The typical dimension will be that of a flux across a boundary, that is, mass per time and area.

It looks as if this is only a change of wording, because streams appear to be closely related to the flux concept. However, this terminology is implicated in the direction of evaluation. In hydrology, the relationships governing the input and output of water for catchments are usually described in terms of matter and energy *fluxes* that can be observed, and the theoretical framework taken from physics relates them to potentials and forces (gradients of potentials) that are not directly observable, but represent the states of the system. These relations are formulated as conservation laws, with mass budgets as the paradigmatic example. Local transport equations are obtained (such as Darcy's Law or Fick's Law) from the conservation laws using variational calculus. It is important, however, that the concept of flux as observed and transport as modeled quantity are formally independent if conservation is not given, as is often the case in nonequilibrium situations (e.g., for rainfall, sedimentation, or chemical weathering).

Streams are a more straightforward abstraction starting from the observed input and output of the ecosystem. Their definition refers to the temporal order among the recorded events and the fact that events can only be produced in an 'online manner,' analogue to that of infinite data types in computation (Gumm 2003). They imply an irreducible diachronic aspect and are therefore retrieved from the memory of an observer rather than being just (synchronic) observations of a state. In the algorithmic approach (Figure 1), states are accessible through observation and imply behavior (here fluxes) by their changes. Therefore, fluxes, when abstracted as a form of behavior, are derived from state changes. This relation becomes reversed under the notion of streams. Documented streams as memorized from past behavior (say a runoff record) imply corresponding internal states (mostly inaccessible to observation). Hence, in this perspective, the states are derived and evaluated from the memorized streams. As long as the stream is noninteractive, the difference between the two approaches is one of perspective only.

The usual conception of fluxes and forces is that they have a deeply rooted translational invariance in time built into them. This is of utmost importance both theoretically as well as culturally in physics: Nonrepeatable experiments and thus nonreproducible results are inacceptable and ignored in the scientific community. We propose that this worldview is impossible for ecosystems. History dependence and the implied uniqueness of each such system are crucial. Unlike observations from purely state-based, memoryless systems that can be reproduced, lost records about streams cannot be substituted in principle. This is reflected by the high value of long-term records in some of the environmental (hydrology) and most of the ecological sciences (Kratz et al. 2003). These facts make the abstraction as streams the more 'natural' for runoff from ecosystems.

We regard the definition and delineation of ecosystems based on the notion of streams as also being the more fundamental one. It allows us to address and to deal with simplicities in ecosystem behavior much more straightforwardly, rather than obscure such simple aspects by using elements of complexity theory from the beginning. Simplicity and universal features in runoff data may, in this view, be regarded as *signatures of internal interaction*. There is by definition no interaction in abiotic streams across ecosystem boundaries. The simplicity of runoff data lies in the fact that choices may produce simple patterns *after* they are made, but that this does not include the ability to predict them (Hauhs et al. 2005).

Whereas streams can be rigorously formalized by coalgebraic notions (see, e.g., Rutten 2000) as primitives of a theoretical approach, fluxes appear in the canonical algebraic approach as secondary (derived) quantities.[9] The direction of formalization becomes reversed in this respect as well: Traditionally, one starts with symmetries in the underlying dynamics (of states) and calculates the resulting order in fluxes, and this then has to be validated from observations. Here, we argue for a data-driven approach in which the order and properties of documented streams are used to reach conclusions on the properties of the underlying interactive process. This approach is unusual for natural sciences but quite common in computer science and engineering; a number of theorems are available for inferences about existence and uniqueness in models.

Interactivity is not a new mechanism that can be constructed (syntactically) by adding additional features to an algorithmic machine. In an algorithmic universe in which interaction does not exist, it cannot be generated *de novo*. However, in a universe of discourse in which we allow for interaction, it can be expressed and demonstrated in the form of interactive computing.

INTERACTIVE STREAMS

The next distinction about streams is whether or not they are generated interactively by the system(s) from which they originate. That is, it concerns the way in which the order in their basic events is implemented: by the action of one system alone or by alternating actions of a system and its environment. The above examples of abiotic streams occurring in the environment of ecosystems and many more such as shortwave radiation, transpiration by vegetation, weathering, or precipitation of secondary minerals in rooting zones are all instances of *noninteractive* streams. The widespread use of such terms in ecosystem research reflects the fact that these noninteractive streams can be recorded much more easily than interactive streams.

One may illustrate this situation with chess playing: In the preparations for a chess game, one can either try to reduce the (seemingly?) interactive situation to a noninteractive one (e.g., find a winning strategy, i.e., solve the game algorithmically, read out the complete chess-related memory content of the opponent[10]). Otherwise, one has to cope with the consequences of interaction (i.e., prepare for the game by training and updating ones own memory with relevant content; increase the ability to *evaluate* a board rather than *predicting* it). One promising strategy to be considered is to reduce the interactivity of the game to a minimum by making the opponent's behavior more predictable. This ideal is implied in the frequently heard advice to novice players "always play the board, not the opponent."

We want to grasp the large gap between scientific and empirical models of ecosystems through this analogy: In the natural sciences, by using one of the two definitions/approaches below, an exo-observer (natural scientist) currently has to avoid interactivity completely if she or he wants to achieve understanding and prediction. Ecosystem managers, however, inevitably have to cope with interaction when they want to sustain a service function. Thus, they will get little help from state models when trying to do so. Even worse, any managerial expertise acquired in the form of heuristics has only a dubious scientific status under the prevailing modeling concept that should be replaced later by some proper understanding of processes and is thus often dismissed by ecosystem researchers (Bocking 2004).

There is no easy classification of whether a data stream across an interface *is* interactive or noninteractive. Furthermore, such a classification may change with time and technical progress. The *a priori* classification of ecosystems into the complexity realm (placing them exclusively under dynamic systems theory) narrows the range of possible models. In addition, it may even narrow the model classes considered to a set of unsolvable tasks. Interactivity may turn out to be an illusion when one ultimately acquires the 'true' dynamic representation of an ecosystem, but we do not care as long as interactive simulations are closer to the nature of the managerial problem than models based on dynamic systems theory.

WHAT IS AN ECOSYSTEM IN NATURAL SCIENCES?

Geosciences

Ecosystems can be delineated spatially on the basis of noninteractive streams at their boundaries. An ecosystem in the perspective of *geosciences* has to fulfil two conditions:

- The smallest region whose boundaries can be characterized completely by noninteractive streams.
- The volume included by this boundary contains some systems that are classified independently[11] as being alive (by unbounded behavioral features such as being able to adapt, evolve, or reproduce).

No further conditions for the internal aspects are imposed.

This first definition imposes the existence of an upper cutoff scale for any biological interaction. This definition has proven useful when investigating the relationship of biotic responses to changes in streams at the boundaries within a larger geochemical and geophysical context, as in 'biogeology' or 'biogeochemistry.' It is (implicitly) widely used in monitoring ecosystem response in the context of environmental changes such as deposition of air pollutants or eutrophication.

Biosciences

The second definition aims at avoiding interaction by heading for the lower cutoff scale of life. An ecosystem can be defined secondly on the basis of the noninteractive components. These are components without any persistent states. Their state is a function of external forces alone. An ecosystem in the perspective of *biosciences* is:

- the largest aggregation of noninteractive components

- that maintains across its outer boundary *unbounded* interaction as an irreducible aspect of its behavioral repertoire with the environment
- while using genes as persistent states for maintaining interaction.

No further conditions for the character of the interactive potential are specified here other than that this type of ecosystem appears as a carrier of some unbounded interactions. This may include phenomena such as an ability to evolve new behavior through (open-ended) evolution. The character of models applied to such ecosystems is in most cases direct (i.e., trying to predict functional output or structural change from given input and initial conditions). The technical challenge in direct modeling is the combinatorial explosion resulting from an iterated combination of the basic (noninteractive) building blocks. In general, such models produce too much data. Their outcome is difficult to evaluate algorithmically. Hence, modelers end up with a severe selection problem that is insufficiently covered by the available data set. Living systems appear as structurally too complex.

A number of biological terms can be made more precise when considering life as an instance of interaction: A multicellular *organism* emerges through a coordinated bounded interaction among locally connected cells (Minelli 2004). A species extends the notion of bounded (ritualized) interaction beyond an organism to an interbreeding population. The bounded set of interactions among organisms that potentially leads to reproduction defines a *species*. The interaction with members of other species remains unbounded. This general relationship among species is characterized as open-ended coevolution – used here as our primary example of an unbounded interaction. Hence, the notion of ecosystems in biosciences addresses the difference between bounded versus unbounded interaction, whereas the notion of ecosystems in geosciences addresses the difference between unbounded and 'amnesic' interaction (i.e., noninteraction across functional boundaries).

A reproducing population of biological agents *is* thus a group of agents using DNA as a hidden persistent *state*. No agent outside this group is able to access this memory in any other way than by observing their phenotype (behavior/structure) or interacting with such phenotypes.[12] The fact that the meaning of the persistent states is hidden from external observers and can only become expressed interactively (with a responsive environment) is the reason for the necessity of the genotype/phenotype distinction in biology. This distinction sets biology apart from other natural sciences, especially in terms of models (Rosen 1991). We regard it as a primary signature of interaction.

The geoscience approach is usually chosen when one studies the functional and spatial embedment of living systems into an abiotic environment: Where does life occur, under which conditions? The bioscience approach is usually chosen when one studies the emergence of living behavior: How did it first arise from its nonliving constituents?

WHAT IS A MANAGED ECOSYSTEM?

We have defined life as an (unbounded) interaction in which the persistence of states is related to genes. The interactive modeling paradigm is accompanied by a typical perspective: An agent typically finds itself embedded *within* an interactive network and views interfaces across which interaction occurs from their 'inside.' An interacting agent may look back at *realized* interactions documented in its accessible memory. The role of an interactive *simulation* is to carry a bounded (and possibly complete) representation of the choices that have been reproducible within this interaction. Then, proper actual choices can be judged by the agent against the choices made during the training phase and their corresponding outcomes in the memorized past.

Memory may exist at an individual or collective level depending on accessibility. One could define the 'self-model' of the agent by individually accessible memory and the identity of a culture by collectively accessible memory, though this is beyond the scope of this paper. Here, we are only interested in a small subset of collective memories: those addressing the realized interactions with ecosystems. By definition (see above), the interaction of humans with co-species can only occur *within* the biotic realm of an ecosystem (disregarding the position of its spatial boundaries to their abiotic environment).

Humans evolved from biotic interactions with co-species. Such interaction was initially unbounded and symmetrical as in any coevolution. With the first human culture, a new form of persistence (besides the persistent states of the genome) and, hence, a new form of memory emerged. Co-species were excluded from this human cultural memory and interaction. Hence interaction of humans with co-species became asymmetrical and bounded for the latter from then onward: Humans *domesticated* other species. Domestication can be viewed as a finite set of intervening options by which further evolution in one species can be stopped or directed by another one. Unlike coevolution, it constitutes an asymmetrical relationship among species. The future survival of domesticated species became dependent on human culture.

The role of humans is unique in being almost the only species that is able to domesticate other species and make their survival dependent on cultural transmission. Note that the two definitions introduced above referred to the unbounded features of life (open evolvability). The existence of domesticated species (and related ecosystems) allows us to introduce another, third notion of an ecosystem that refers to bounded interaction. In the case that one succeeds in establishing an ecosystem in which a domesticated species becomes a dominating population (e.g., a field of wheat, or pasture with a herd of cattle), the concept of bounded interaction transfers to the ecosystem. Choice options open for the domesticated species are culturally constrained to a finite set that is exhaustively known to the domesticating human culture. Ecosystem management becomes a mixture of functional and interactive relationships between humans and the system hosting domesticated species. Abiotic streams across ecosystem boundaries, as defined above, are examples in which the functional relationship is appropriate: for example, watering or supplying additional nutrients to an ecosystem. The selection of individuals for breeding is an example in which a bounded interactive relationship holds.

Experts in farming, pasture, forestry, and so forth use the term ecosystem in this third type of meaning. It is the system that they can interactively force into a standardized overall function (Bocking 2004). This interaction is *bounded*; it can be represented in a tradition and can be applied in a sustainable manner. That is why it can be embedded into an external overall function for the civilization of which it is a part: providing timber, fiber, food, and so forth for the society.

The dilemma of the two scientific notions above for an ecosystem is a methodological one. As long as observers remain 'exo-' with respect to the observed system, they will encounter instances of unbounded interaction that occur within it. We conjecture that these phenomena (and the corresponding states) can neither be identified by inverse modeling (geosciences) nor can they be generated *de novo* by direct modeling (biosciences). The rigorous scientist, who strives to study untouched or only experimentally conditioned ecosystems, may not be able to get rid of internal unbounded interaction. This feature will limit and intervene in any predictive modeling attempt. If interaction is taken seriously, the difficulties discussed above appear as a signature of a principal limit rather than as technical difficulties to be overcome by refinements in measurements or more comprehensive modeling attempts.

Successful managers of an ecosystem are able to interactively prevent unbounded choice within the system occurring. If they can demonstrate reproducibility, they have achieved the ultimate management condition: sustainable ecosystem management. The 'price' that has to be paid for this, however, is that managers are inevitably participatory endo-observers for some interactions. Their observations and memories are by no means objective. Here, (interactive) simulation may be a decisive new technology that helps to document, investigate, and disseminate such expert knowledge beyond the idiosyncracies of its origins. The flight simulator example can in this respect be extended into ecosystem management (Hauhs et al. 2003).

A model comprehensively representing a bounded interaction may serve at the same time as the carrier of norms for proper intervention. It may become a basis for the evaluation of new instances of interactions in a similar manner as a minimal (explanatory) model may be the basis of predictions in the algorithmic paradigm. If a chess computer can represent and handle anything (bounded) that might happen in chess, it may also direct a novice to proper moves; if a flight simulator includes anything (bounded) that can happen to the pilot of a specific plane, it can be used for training. If a forest growth simulator includes what has happened in a particular type of forest, it can be used to train thinning operations. In real-world situations, function and interaction may thus occur in a nested manner making them difficult to separate. Ecosystem utilization is an example in which a bounded interaction can be delegated to experts such that the whole system serves a function such as providing food. Note that the embedding relationship between interactive and functional aspects can also occur in a reversed manner. A musical instrument or a lasso (Ingold 1994) are both examples in which a functional tool with complete physical (algorithmic) description can be used by experts to serve an interaction. It can also only be learned interactively.

In interactive computing, simulations do not provide an explanation of what has happened; they do not represent the 'natural laws' governing the true dynamics as in the case of algorithmic models. However, they represent the past choices that may reoccur in a bounded interaction and may hence represent social norms in an intersubjective and novel way. This makes it possible to evaluate rare and decisive situations in a systematic way (e.g., in chess, the aviation industry or forestry).

Table 1. Limits of the two modeling paradigms and how they are related to prominent simulation tasks in science, engineering, and ecosystem research

	Algorithmic models	**Interactive models**
Accomplished tasks	Many examples in physics Reconstructing plant structure (L-Grammar)	Chess computer Flight simulator for pilots
Current limit	Weather prediction	Flight simulator for foresters Assessment of empirical models (e.g., in hydrology)
Out of reach for technical reasons	Assessment of empirical models Predicting forest growth under climate change Open-ended evolution	Flight simulator for nature conservationists
Out of reach in principle	Nothing of practical relevance?	Predicting forest growth under climate change Open-ended evolution 'Flight simulator for God'

The new model type may lead to a very different perception of where to expect technical and fundamental limits (Figure 1; Table 1). Some problems such as predicting an ecosystem response under an altered climate, appeared to be complex but solvable in principle under the algorithmic modeling paradigm: These will be reclassified in the new paradigm, and may become unsolvable in principle. Despite the disappointment that such a result may mean for ongoing research projects (e.g., climate change), in the long term, we consider it a step forward. If a situation is truly interactive in the sense we have used the term here, there is no way to substitute for missing experiences from an open interaction (e.g., if a type of choice or behavior has not occurred yet). This is due to a variant of the combinatorial explosion mentioned already: the mismatch between genotypic potential and actual phenotypic expression. Only an almost negligible fraction of the former can be realized within the lifetime of an organism, an ecosystem, or even a whole species.[13] On the other hand, other problems such as assessing empirical knowledge and expertise in ecosystem management appeared as similarly complex under the old paradigm. Under the new paradigm, these problems may now be within the reach of modern IT.

A different case is nature conservation, a problem that, even with today's IT, may remain technically too hard for interactive simulation models (see Table 1). In these cases, the goal is to keep an ecosystem-wide set of species, of growth potential, or its biodiversity. We still do not know whether this is a management task that can be organized as in forestry or agriculture, or a goal of a different character (e.g., in terms of ethical values). If it becomes a management task, then any *reproducible* measure of success has to be based on *bounded* interaction with technical norms. In other words, successfully managed ecosystems will, under the goal of nature conservation, ultimately become domesticated by halting open-ended evolution. If not, the result of any protective action cannot be judged by its *results*, but rather by its *intentions*. Intent, however, is not an operational criterion for unbounded interactions among different species.

Ecosystems not touched by humans (if there are any), *natura naturans*, cannot be represented by a bounded interactive simulation. Modeling and evaluating their behavior remains elusive under the second modeling paradigm.[14] The anomaly mentioned in the introduction between what should be and what appears to be possible with respect to ecosystem management is thus resolved. The addition of human goals and proper interventions is what makes the modeling problem tractable for *natura naturata* under the new interactive paradigm, whereas human interference has been considered as a disturbance rendering modeling even more difficult under the traditional algorithmic simulation approach.

CONCLUSIONS

Up to now, most simulation models developed in the social and biological sciences still use the algorithmic modeling paradigm. Technically, such models do not leave the realm of dynamic systems theory. These approaches abstract from any interactive aspects of the modeled system. The results are ambivalent and leave ecological (and, as far as we can see, also social) modeling in a dilemma: The models still do not yet deliver contra-intuitive predictions relevant for management. However, these models are very useful when used as communication tools for arguing about the cases studied (Bousquet and Le Page 2004). A model developed for prediction under dynamic systems theory becomes a communication tool when it fails to predict blindly and is thus calibrated to observations. For this purpose of communicating and documenting existing experiences, however, more efficient, interactive simulation tools are available today. As long as various forms of interaction are not defined and studied more rigorously, simulation models may focus on the wrong aspects of ecosystems. While we may still fail to predict ecosystems, we miss a chance of improving their evaluation by experts.

The interactive modeling paradigm provides us with a different and new way of representing human knowledge. It may become recognized as a dual form of the traditional approach. The proliferation of artificial objects resulting from functional and industrial production interested philosophers in the nineteenth century. Today's increase in interactive simulations of artificial (choice) behavior is facilitating communication and sets current technology changes against this historical background of facilitating production. (Interactive) simulation attracts the attention of philosophers of science as documented by this book.

Interaction cannot be created *de novo* in a computer. However, if it is a useful concept, with a promising potential to become a formal rigorous one, some of its existing bounded forms can be *transferred* to a computer. It allows us to abstract from and to deal with a human perspective toward valuation and choice from an endo-perspective. This time, it is not an abstraction with respect to observation (and spatial perspective) as in Renaissance times, but has a memory and a temporal perspective. It may provide areas of professional expertise in ecosystem management with an intersubjective way of documenting and communicating knowledge about bounded sets and the proper order of decisions in certain interactive situations.

In a more general sense, a better understanding of simulation technologies and especially interactive ones may acquire a role comparable with the mastering of perspective in arts. For visual perception, the invention of perspective became a historical stepping stone for the 'cognitive enlightenment' and subsequently modern science. Today's interactive simulation is about to acquire the technical potential for rehabilitating memory (alongside vision) as a similarly reliable source of intersubjective knowledge. In some restricted areas such as chess or flight simulation, it has become a carrier of norms and is already used routinely for training to expert levels. If this approach can be extended to the human relationship with ecosystems, it may trigger a corresponding 'normative enlightenment.'

We have argued above that under this modeling paradigm, it becomes easy and straightforward to define the key notions of ecosystem research and modeling. In addition, it allows us to discuss technical and principal limits, and this seems to give a simple explanation of past successes and failures in ecological modeling. It is at least a complementary approach to the same systems from a distinctively different perspective, and exploits knowledge on them that was dismissed all too quickly in the traditional approach.

Taken together, this seems to be sufficient reason for the distinction introduced above and especially interactive simulation to be taken more seriously by philosophers of science as well. Which modeling tools will yield better results in ecology and ecosystem research will, of course, depend on empirical testing.

*University of Bayreuth, Germany
**Norwegian Institute for Forest Research, Ås, Norway

NOTES

1 "The central concept of Newtonian mechanics, from which all others flow as corollaries or collaterals, is the concept of *state*, ..." (Rosen 1991).

2 If ecosystems are viewed under dynamic systems theory, one could ask instead: ... *despite* continued human interference. However, the search for dynamic models assessing the stability of ecosystems is an old and still controversial field, which we shall not go into here (McCann 2000).

[3] We use the notion of 'skill' in the same sense as Ingold (2000).

[4] Nature conservation may be regarded as an exception. There the management goal is often to sustain an ongoing interaction with 'nature' (see last section).

[5] The presence of noise is unrelated to interactivity. Noise does not make choices by definition.

[6] Representations are often modified with the effect that a model is largely simplified or extended beyond its 'grounding in reality.' In the first case, the model seeks the most concise representation of a state; in the second, the most comprehensive representation of a choice.

[7] However, not every computer has to be an algorithmic machine. There may be aspects in the heuristics of IT engineering that have not yet been properly formalized. Interactive computing can be considered as an example; it is economically important, but not all aspects of it have been given a formal grounding (see Introduction of Turi [1996]).

[8] As used in the computer sciences, particularly in coalgebraic approaches to computational structures.

[9] We tried a formalization of fluxes as boundary-determining objects for ecosystems earlier (Hauhs and Lange 1996). However, this does not lead to any fundamental and concise notion as is possible for streams in coalgebra.

[10] Try this only if it is a computer!

[11] The logic relationships among the attributes usually used to define life remain unclear (Ruiz-Mirazo et al. 2004).

[12] There may be observations of the genotype, but since the mapping from genotype to phenotype is an interactive one, this mapping cannot be identified by observations alone (but requires an interactive approach itself). This leads to the conjecture that the proteomic research program must fail on principal grounds similar to the alchemistic program, that is, failure due to choosing a wrong model category (and hence ignoring limits).

[13] With the possible exception of unicellular organisms.

[14] Whereas explaining their behavior under the first paradigm only appeared to be very difficult.

REFERENCES

Arbab, F. (2005). "Abstract behavior types: A foundation model for components and their composition", *Science of Computer Programming*, **55**: 3–52.

Beven, K. (2001). "On modeling as collective intelligence", *Hydrological Processes*, **15**: 2205–2207.

Bjerkenes, V. (1904). "Das Problem der Wettervorhersage, betrachtet vom Standpunkt der Physik und Mechanik", *Meteorologische Zeitschrift*.

Bocking, S. (2004). *Nature's Experts. Science, Politics, and the Environment,* New Brunswick, NJ: Rutgers University Press.

Bousquet, F. and C. Le Page (2004). "Multi-agent simulations and ecosystem management: A review", *Ecological Modelling*, **176**: 313–332.

Brooks, R. (2001). "The relationship between matter and life", *Nature*, **409**: 409–411.

Goldin, D., S.A. Smolka, P. Attie, and E. Sonderegger (2004). "Turing machines, transition systems, and interaction", *Information and Computation Journal*, **194**: 101–128.

Gumm, H.P. (2003). "Universelle Coalgebra", in T. Ihringer (ed.), *Allgemeine Algebra*, Lemgo, Germany: Heldermann Verlag, Appendix.

Hauhs, M. and H. Lange (1996). "Das Problem der Prozeßidentifikation in Waldökosystemen am Beispiel Wassertransport", *IHI-Schriften Zittau*, **2**: 212–222.

Hauhs, M., F.-J. Knauft, and H. Lange (2003). "Algorithmic and interactive approaches to stand growth modeling". in A. Amaro and D. Reed (eds.), *Modeling Forest Systems*, Wallingford, UK: CABI Publishing, pp. 51–62.

Hauhs, M., J. Koch, and H. Lange (2005). "Comparison of time series from ecosystems and an artificial multi-agent network based on complexity measures", in J.T. Kim (ed.), *Systems Biology Workshop* at the VIIIth European Conference on Artificial Life (*ECAL 2005)*, Canterbury, Kent, UK: University of Kent, 12pp.

Ingold, T. (1994). *Companion Encyclopedia of Anthropology*, London: Routledge.

Ingold, T. (2000). *The Perception of the Environment – Essays in Livelihood, Dwelling and Skill*, London: Routledge.

Jakeman, A.J. and G.M. Hornberger (1993). "How much complexity is warranted in a rainfall-runoff model", *Water Resources Research*, **29**: 2637–2649.
Kimmins, H., C. Welham, B. Seely, M. Meitner, R. Rob, and S. Tom (2005). "Science in forestry: Why does it sometimes disappoint or even fail us?", *IUFRO 205*.
Kratz, T.K., L.A. Deegan, M.E. Harmon, and W.K. Lauenroth (2003). "Ecological variability in space and time: Insights gained from the US LTER Program", *Bioscience*, **53** (1): 57–67.
Lansing, J.S., J.N. Kremer, and B. Smuts (1998). "System-dependent selection, ecological feedback, and the emergence of functional structure in ecosystems", *Journal of Theoretical Biology*, **192**: 377–391.
McCann, K.S. (2000). "The diversity-stability debate", *Nature*, **405**: 228–233.
Minelli, A. (2004). *The Development of Animal Form - Ontogeny, Morphology and Evolution*, Cambridge, UK: Cambridge University Press.
Peters, R.H. (1991). *A Critique for Ecology*, Cambridge, UK: Cambridge University Press.
Pittroff, W. and E.K. Pedersen (2005). *Ecological Modeling, Encyclopedia of Life Sciences* (http://www.els.net), Chicester, UK: Wiley; doi:10.1038/npg.els.0003270.
Rosen, R. (1991). *Life Itself – A Comprehensive Inquire into the Nature, Origin, and Fabrication of Life*, New York: Columbia University Press.
Rubin, D.C. (ed.), (1995). *Remembering Our Past – Studies in Autobiograhical Memory*, Cambridge, UK: Cambridge University Press.
Ruiz-Mirazo, K., J. Peretó, and A. Moreno (2004). "A universal definition of life: Autonomy and open-ended evolution", *Origins of Life and Evolution of the Biosphere*, **34**: 323–346.
Rutten, J. (2000). "Universal coalgebra: A theory of systems", *Theoretical Computer Science*, **249** (1): 3–80.
Schellnhuber, H.J. and V. Wenzel (eds.). *Earth System Analysis*, Berlin, Heidelberg, New York: Springer.
Schütz, J.-P. (2001). *Der Plenterwald und weitere Formen strukturierter und gemischter Wälder*, Berlin: Parey Buchverlag.
Turi, D. (1996). *Functional Operational Semantics and its Denotational Dual*, Amsterdam: Free University of Amsterdam.
Ulanowicz, R.E. (2004). "On the nature of ecodynamics", *Ecological Complexity*, **1** (4): 341–354.
Wegner, P. and D. Goldin (1999). "Interaction as a framework for modeling", in P. Chen, J. Akoka, H. Kangassalo, and B.Thalheim (eds.), *Conceptual Modeling: Current Issues and Future. Lecture Notes in Computer Science 1565*, Springer, Heidelberg, pp. 243–257.
West, G.B. and J.H. Brown (2004). "Life's universal scaling laws", *Physics Today*, **57** (9): 56.

CHAPTER 5

DON IHDE

MODELS, MODELS EVERYWHERE

The newest technological toy, the 'computer,' has given us a complexity machine with which ever higher degrees of complex phenomena can be computed, manipulated, and produced in a variety of imaging forms: charts, graphs, simulations, and images. Indeed, it may be or be becoming the twenty-first century's *epistemology engine.*

I call an epistemology engine, some technology which then is used to model the process of knowledge production. Previously I have argued that the *camera obscura* precisely served that role in *early* modern philosophy. It was explicitly used as a model of knowledge by both René Descartes in the *Dioptrics,* and even more explicitly by John Locke in the *Essay on Human Understanding.* While I shall not retrace that analysis here, I tried to show how the subject/object; external/internal; and knowledge as representational all follow from the way the *camera obscura* worked according to seventeenth century understanding. Whether or not one should ever take a technological model for knowledge production or human understanding aside, I have tried to show that the *camera* model is now outdated since it no longer models the kind of practices which produce contemporary styles of knowledge.

By very *late* modernity, a few philosophers have partially identified computation devices and processes as such an epistemology engine, but no one to my knowledge has done so with as much positivity as Descartes and Locke did with the earlier toy. Hilary Putnam has flirted with the idea that the computer serves this role today, and a loose fit borders on more than the metaphorical amongst 'computational models of mind' analytic philosophers. These philosophers do think that there is more than a brain-computer metaphor, as indicated by the wide use of "brains-in-vats" by Daniel Dennett in his *Brainstorms* (1978). But this engine has not yet stuck as fully as the 'theatre of the mind' camera.

So, before deciding how suggestive computational devices may be for epistemology, let us look at some of the main features which the 'computer' as a complexity device can do:

- First, computation can perform massive computations at speed; calculations which would take hundreds of mathematicians decades of time, can now be done

J. Lenhard, G. Küppers, and T. Shinn (eds.), Simulation: Pragmatic Construction of Reality, 79–86.

in manageable, finite periods of time. As a speedy calculation device, able to handle very complex calculations, computation gives us a new amplification of one human capacity never before possible.

- Moreover, the computations possible can deal with many multi-variables, which, in turn, can then be graphed. In turn, modeling via graph frequently reveals unsuspected patterns – for example, a multi-variable graphing process some years ago was applied to the levels of lead in the atmosphere, correlated with human activities from antiquity to the present. The result is one which clearly depicts the role of homogenic activity upon atmospheric phenomena.

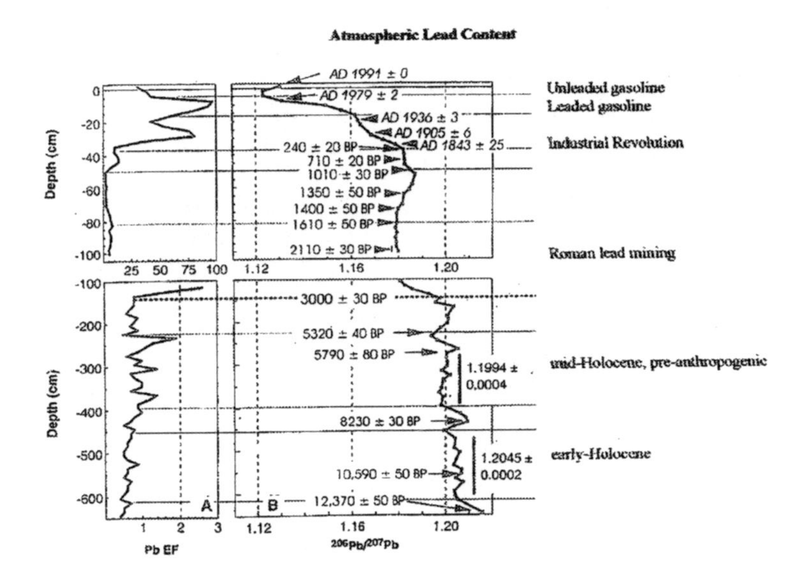

Figure 1. Atmospheric Lead Content. (Reprinted with permission from Shotyk et al., Science ***281****: 1635–1640, 11 September 1998. Copyright 1998 AAAS.)*

- Better still, computer modeling can move from data-to-image-to-data. By using algorithms, one can produce images which are 'readable' at a glance, or, one can reduce images to data for analytic purposes. Peter Galison's *Image and Logic* (1997) showed how this capacity in physics simulations, tended to give the edge to imaging processes in late twentieth century physics. The *Brookhaven National Laboratory*, in producing a poster advertising its heavy ion accelerator, shows both data-graph (logic) and simulation image (image) detectors.

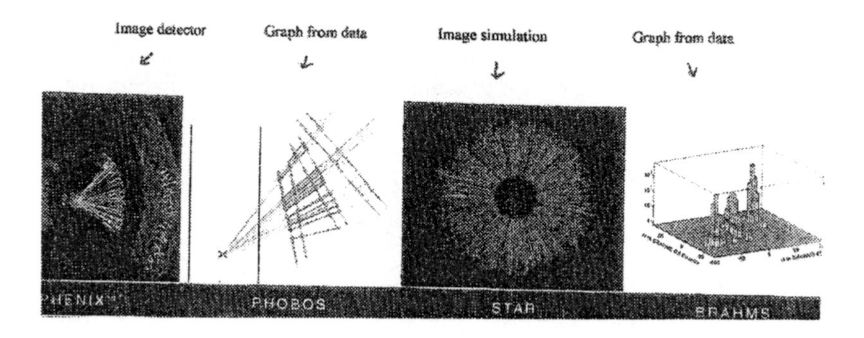

Figure 2. Relativistic Heavy Ion Collider Detectors

- The data-image reversibility has also allowed for modeling of long term processes not previously manageable. My colleagues, Pat Grim and Gary Mar, in *The Philosophical Computer* (1998), were able to model semantic paradoxes in three-dimensional projections in such a way that interesting differences were shown to obtain between different types of paradoxes.

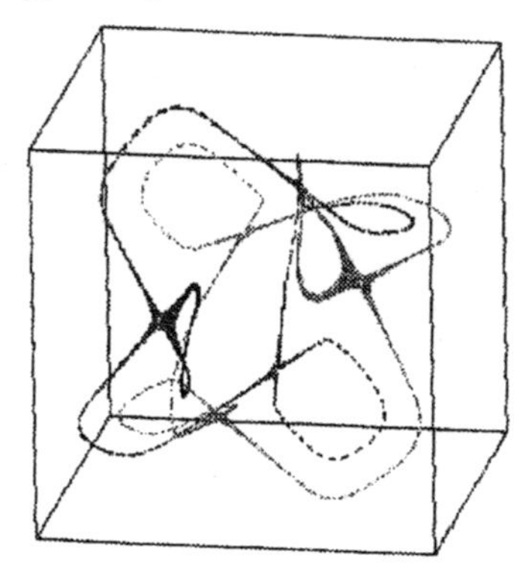

Figure 3. Strange attractor semantic paradox. (Permission by Patrick Grim.)

- One result of this – contrary to what is sometimes thought of as a computational move toward 'disembodiment' – is a return of critical interpretive activity to the humanly *perceivable* through images. Images produced by computations thus produce what can be seen at a glance, thus engaging the visual gestalt capacities of embodied humans.

Those of us familiar with the models which such processes produce recognize that images are, in effect, mediations. Here, I shall concentrate upon the imaging processes employed in models, but with a particular concern. Early modern epistemology was an epistemology centered in *representations*. In their simplest forms, representations were what I call *isomorphic images,* that is, images which are 'like' that to which the images refer. The actual optical model, employed by both Descartes and Locke, was the *camera obscura,* for which the images – which stood for the impressions or sensations in the mind – to be 'true' had to be isomorphic.

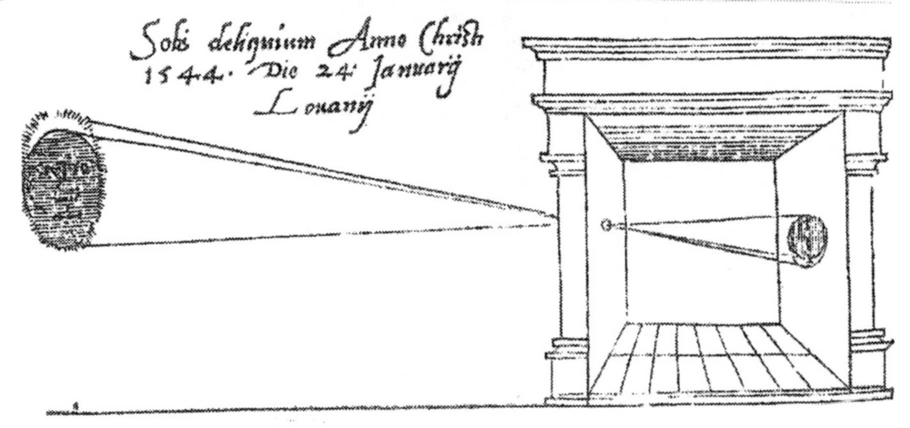

Figure 4. Sixteenth century camera obscura

Thus, the imaged Sun projected upon the back wall of the camera is 'like' the external Sun in shape, configuration, and so forth. Except it is not! The imaged Sun is flat or two-dimensioned; it is inverted or upside down; and in any actual *camera obscura,* does not even approximate the intensity of light of the external Sun. In short, the *camera obscura* radically transforms the Sun into the imaged Sun. Similarly, in a later modification of the *camera,* that is the nineteenth century *photographic camera,* the process of producing the image is one which 'fixes' the image of the Sun, yet another transformation.

Now, so long as that which is imaged is available both to direct perception and the mediated perception provided by the image, one can compare Sun with imaged Sun. And, when this is done, one can also see that in definite and limited senses, the imaged Sun has certain advantages because of the technological transformation entailed. Looking at either an *obscura* or *photographic Sun* will not make you blind! You can also return again and again to the image to take notice of features perhaps not noted at first glance. (But you can also take account of features which make the imaged Sun different from the perceived Sun: size, two-dimensionality, stasis, etc.) But note, all this is possible *only if one has the comparative capacity to differentiate between the image and the object imaged.*

With contemporary imaging of the sort for interest here, that comparative capacity simply does not exist. I shall take as my example what I call 'whole earth measurements' or simulations related to global warming. Here are some recent such simulations produced as images of the whole earth.

In a strict sense, these are not 'images' in the previous sense of simple, isomorphic depictions of a perceptible object. Rather, these are graphic depictions of phenomena which *could not and cannot* be perceived from an embodied and situated perspective, not even one from a satellite perspective. The schema which is depicted

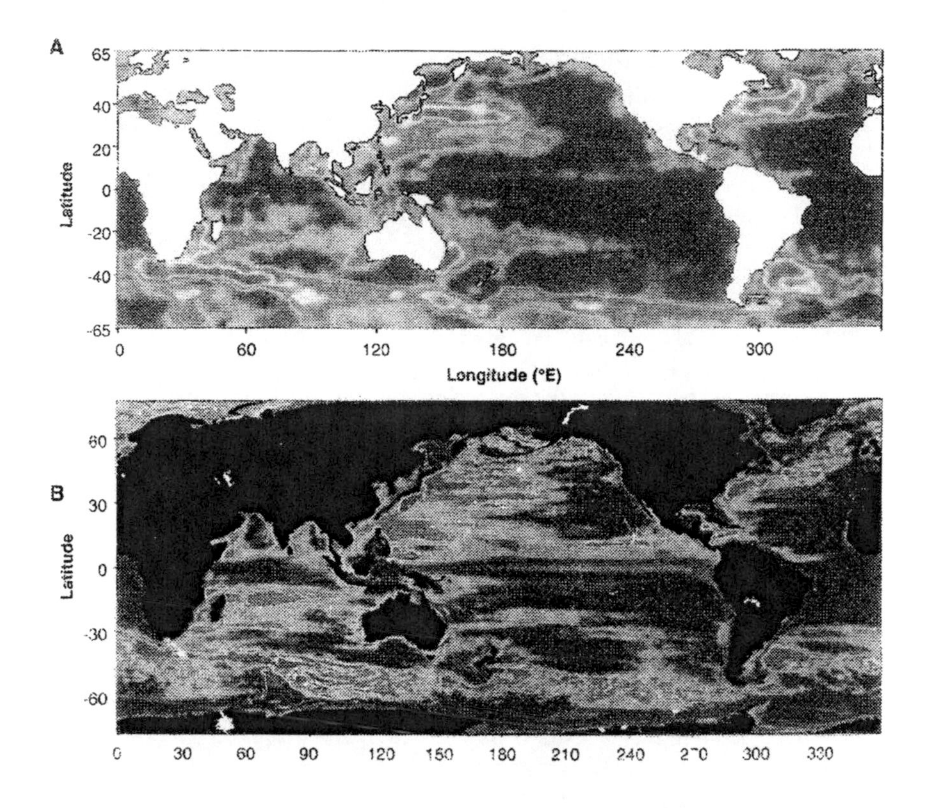

Figure 5. Simulations related to global warming

is a ‘whole earth’ projection, reduced to a single image. It is also depicted with a color convention, in the original, using ‘false color’ to map intensities – and unless one is privy to this convention, it is not even possible to know what is being depicted. In this case it is ocean levels, not, for example, heat ranges which could also be so mapped. It also incorporates an older convention, with the arctic circle at the top; the antarctic at the bottom (a convention which reversed older conventions). In short, this is not really a ‘picture’ or what we might ordinarily think of as an ‘image.’ It is much more like a ‘map’ in a special sense.

It does have vestigial isomorphic features, one is familiar with the very high perspective which produces continental shapes (illustration from satellite view of the earth). These could be seen were one on the space station, but here are reduced to a flat projection with its built in distortions, and including a 360 degree sweep. All of this and much more is built into this model. Yet, to the informed perceiver-‘reader’ of this depiction, all of this is available at a glance for an ‘aha’ recognition – “so that’s how much the oceans have risen!” But, however seen or read, one cannot simply compare the ‘knowledge’ produced by the model with ‘reality’ since one has never had the ‘reality’ of the whole global view!

Let us complicate the scene a little more by introducing dynamics. We can get increasingly sophisticated results about 'real' earth history by making more and more things 'speak' or give us measurements from the past. Greenland ice cores, subject to ion analysis, can take us back several tens of thousands of years; the same for Antarctic cores. Ocean bottom sediments can also reveal patterns which yield ancient temperatures. Put all these together and one gets a piecemeal mapping, not as 'coherent' as our 'image' has it, but perhaps in each piece a bit of accuracy. But, strictly speaking, there remains no way to compare 'real' earth history and the simulated 'history.' Yet, this language continues to pervade much discussion – and much objection to the simulator's claims. This is, however, to assume that simulations are in some sense 'representations.' I do not think they are, although I allow that the vestigial isomorphism suggested by the way the 'scientific' image is presented tempts one to believe that. Rather, I do not think contemporary science imaging is either a 'picture' or a 'text,' although it probably does have map-like features. These, in turn, presume skills of interpretation which map readers must have learned.

The map is never the territory. Borges' fictional Chinese Emperor who needed a paper larger than the territory in order to map the territory is one such absurdity noted. Rather, imaging in the context of simulation and modeling is more analogous to a critical, interpretive *instrument,* through which we see and read. Insofar as a simulation 'images,' it does not do so on the basis of any copies or isomorphic representations since it is nothing like either the optical lens systems of microscopes or telescopes, nor of a *camera obscura* and the progeny therefrom. *There is no original from which to copy.* Yet the end result *is* image-like; it is a gestalted pattern which is recognizable, although it is a *constructed* image.

I want now to examine a few features of what is really a relatively new critical-interpretive instrument. The literature about the uses of models and simulations remains rife with representationalist language. "How closely does the model match the real?" But, we don't have the real separately to tell if there is a match or not. Rather, in one sense, it is the instrument, the model, which gives us the 'real.' Or, at best, if we have a 'real' record of some sub-pattern, perhaps then we can say the model does match a sub-pattern. What we are after, however, is a depiction of a lot of composite features which we may have separately. It strikes me that what we have is an analog to the learning of tool use familiar in much earlier science. As Andrew Pickering points out, there is a lot of tuning and skill to attain before the instrument becomes as transparent as it can be (Pickering 1995). And, we have to learn how to distinguish 'real phenomena' from 'instrumental artifacts.' Double images, 'auras' or 'halos' frequently bugged early telescopy in analogy to many model 'artifacts' in simulations.

Return to my earlier whole earth chart: What the simulation image depicts is a very complex *composite* of multiple measurement instruments. Ocean buoys, satellite readings, deep sea probes, and a wide variety of separate measurements are tomographically combined to make the image shown. If one has enough variables, has tinkered well enough, then one hopes the image is 'adequate.' I previously claimed that this kind of image is more like a map than a picture.

Figure 6. Sea floor image

Here is one recent *map* example which is also constructed and composite. These are recent ocean bottom maps which drew from Cold War techniques originally produced for finding and/or hiding submarines.
The construction process utilizes a series of different technologies:

- Gross features come from averages of many satellite passes which image ocean surfaces, in turn analyzed via gravitational effects to show sea mounts and such;
- then, with multi-side scan radar, more detail emerges;
- and, finally, where fine detail is needed, a photographic and optical scan can be made.

But, once again, it is the tomographical capacity of computation which combines and constructs what in this case is the 3-d projection of ocean bottom map. This example, while showing the constructive and composite features, would for the previous example be only one of an even wider set of variables.

What I have been illustrating is not only a set of some of the most expensive 'pictures' ever produced, but the way in which 'models, models, everywhere' is taking hold. Contemporary imaging is 'constructed' imaging. When compared with early artistic use of the *camera obscura,* for example, the tracing of the inverted image, while 'active,' was drawing-by-the-lines. Photography, as a later adaptation of the *camera,* was in a special sense 'passive' in that the chemical process did the 'drawing.' Today's constructed imaging retains an analogue to art processes, in that the result is well-planned, laid out with results in mind, and thus more active than the seeming 'photo-realism' of earlier forms of science imaging.

Stony Brook University, New York, USA

REFERENCES

Dennett, D. (1978). *Brainstorms: Philosophical Essays on Mind and Psychology*, Cambridge, MA: MIT Press/Bradford Books.

Galison, P.L. (1997). *Image and Logic: A Material Culture of Microphysics,* Chicago, IL: University of Chicago Press.

Grim, P., G. Mar, and P. St. Denis (1998). *The Philosophical Computer. Exploratory Essays in Philosophical Computer Modeling*, Cambridge, MA: MIT Press/Bradford Books.

Pickering, A. (1995). *The Mangle of Practice: time, Agency and Science*, Chicago, IL: University of Chicago Press.

LAYERS OF INTEGRATION

CHAPTER 6

GÜNTER KÜPPERS AND JOHANNES LENHARD

FROM HIERARCHICAL TO NETWORK-LIKE INTEGRATION: A REVOLUTION OF MODELING STYLE IN COMPUTER-SIMULATION

INTRODUCTION

In general, models in science are highly idealized and ignore most of the effects dominating reality. A prominent example is the theory of ideal fluids that neglects the effect of dissipation. Hence, strictly speaking, models in science are unrealistic. They represent an ideal world that is believed to lie behind the diversity of phenomena in the real world. To obtain more realistic models – in our example, the theory of real fluids – additional effects (e.g., the viscosity of the fluid) have to be integrated into the basic model. An adequate choice of effects depends upon the purpose for which a model is built. For example, if one tries to understand the phenomenon of hydrodynamic convection patterns, because of the relevance of dissipation basic principles of thermodynamics must be added to the mechanical equations of hydrodynamics. In this case, integrating new effects into the basic model of ideal hydrodynamics is not a problem, because this takes place under the uniform paradigm of physics. Hence, the integration of new effects is based on reliable theoretical grounds.

In most fields of practice, however, the process of applying science has to face serious problems: All kinds of effects must be taken into consideration in order to gain the relevant knowledge to tackle real-world problems. In this case, integration is no longer possible on the grounds of a common theoretical conception. Various scientific disciplines may become involved, contributing heterogeneous models; questions of instrumentation and technology may arise; and there may well be problems of political regulation, social acceptance, and economic success. These problem areas have to be integrated into an overall strategy of knowledge production without a common theoretical ground and even without the leading role of science. This new form of integration may be called pragmatic integration, because it has to be successful but by no means correct. Therefore, from the opposing perspective, this pragmatic integration can be seen as a fingerprint of application-dominated knowledge production in different problem areas in society.

J. Lenhard, G. Küppers, and T. Shinn (eds.), Simulation: Pragmatic Construction of Reality, 89–106.

Pragmatic forms of integration challenge knowledge production in several fields. There are no general methods for making integration a success. Because of the lack of theoretical paradigms, it must be determined by technical or social constructions: as a kind of plumbing with respect to the methods and as a social networking with respect to its social practice.

In the last decades, computer simulations have become established as a powerful instrument in science and technology for the theoretical solution of complex problems, especially the dynamics of complex systems. Simulation models – traditional ones like differential (difference) equations as well as phenomenological ones like agent-based models – are used as a kind of generative mechanism to imitate the dynamic of a complex system. For this reason, computer simulations are more than classic models in science. They are complex algorithms that open up the possibility of running theoretical models as computer programs on a computer in order to show the internal dynamic behavior of the models.

Because, nowadays, computer simulations are used in different fields of practice, the problem of integrating different contexts of application is a current problem within simulation. What is a realistic simulation of a complex behavior? Is it, for instance, realistic because it uses the basic equations of physics, or is it realistic because all important effects are integrated beyond the theoretical paradigms of these effects? And, if so, what are the important effects? Which ones may be neglected for the sake of simplicity, and which ones not?

Many simulationists argue that the reality approach of simulations cannot be decided on the basis of the quality of the underlying models. On the contrary, it must be decided on the basis of the quality of the result. In other words, a realistic simulation is a simulation that is believed to be realistic. This transition to a new approach toward realism within computer simulation can be demonstrated in the case of climate research. Climate simulation models are, at present, the most complex ones, running only on the biggest computers in the world, involving also high levels of commitment to climate policy. Because of their political context, these simulation models must be realistic and reliable at the same time. This need has fueled a revolution in modeling style – a transition from a hierarchical to a network-like integration of models.

In the following pages, this revolution will be illustrated in two steps: We shall start with the development of hierarchical integration and the essential breakthroughs for the simulation method (1955 to mid-1990s) and then go on to analyze the shift toward a network-like architecture of simulations in climate research driven by the political demands for integration.

SOLVING UNSOLVABLE PROBLEMS

The Simultaneous Birth of the Electronic Computer and the Simulation Method

The development of both the electronic computer and the simulation method took place nearly simultaneously. The time and place of the latter's birth can be located at the end of World War II in Los Alamos, with a couple of applied mathematicians working together in the Manhattan project acting as virtual parents. One of the problems that turned out to be crucial for the development of the atomic bomb was the diffusion problem of neutrons. Although the physical principles had been known for a long time, the underlying equations proved impossible to solve. The electronic calculation machine developed at this time opened up new possibilities for the treatment of such complex problems. However, to use the new electronic machine, new methods of dealing with mathematical problems had to be developed: the Monte Carlo method, modeling via cellular automata, as well as finite difference approaches. All these methods can be seen as an attempt not to solve a system of complex mathematical equations but to imitate the dynamic encoded in the set of equations. These new approaches to complex problems constitute ways to use computer simulations as scientific instruments. The Polish mathematician Stanislaw Ulam played a key role in the invention of a couple of these methods – if one wants to mention a central figure in the invention, or construction, of simulation methods, he is the one.

The Monte Carlo method may serve as an example. This method goes back to the joint effort of Stanislaw Ulam and John von Neumann when working together on the Manhattan project in Los Alamos. Monte Carlo may count as the first simulation method.[1] Because the neutron diffusion problem was unsolvable by analytical methods and because a lot of experimental data were available on the scattering of individual neutrons by atoms, they were looking for a method by which they could obtain the behavior of a macroscopic neutron beam from these individual scattering events. Instead of calculating a solution of the basic equations, a statistical method was employed to imitate the behavior of diffusion. The following example will illustrate this approach.

Imagine that you intend to determine the volume of a certain body via Monte Carlo. You can embed the body into a cube with a known volume. The surface of the body defines an analytical function whose integration would give the so-called primitive. In many cases, this analytical approach is impossible, and the primitive cannot be calculated. The idea is to replace the (unknown) primitive by a ratio that can be determined 'empirically,' or quasi-empirically, by iterating computer runs. The computer determines a point within the cube at random. If this point belongs to the body, the trial is said to be successful. By re-iterating this random choice, one can determine the unknown volume as the ratio of successful trials out of a large number of trials. In other words, the surface function is not integrated numerically. Instead, this process is imitated by a generative mechanism.

During his work in the context of the Manhattan project, von Neumann tackled problems like the propagation of shock waves, another problem that could not be treated with analytical methods. This meant they could not be treated mathematically at all. The important point is that the relevant laws of hydrodynamics are very well known. They are expressed in a system of nonlinear partial differential equations (PDEs), whose solution determines essential properties of the behavior of the system under investigation. But to solve such a system of equations, one would have to find a set of analytical functions satisfying the set of differential equations that make up the integration in the technical mathematical sense. This is (in most cases) impossible for a set of complex nonlinear equations. Computer simulations changed the situation fundamentally – a new strategy for solving this problem had become available.

First, the set of PDEs is replaced by so-called finite difference equations (FDEs). Space and time are endowed with a grid structure reflecting the limited capabilities of a computer – it can only handle discrete objects. The FDEs are calculated at the grid points and evolve step by step over time. Just imagine a kind of approximation: If the grid becomes finer and finer, the FDEs will become identical with the PDEs – at the limit. The point, however, is that this holds only in principle. There is not only a limitation of computer time, which does not allow infinite small grids, but also a problem of truncation errors, because the calculation within the grid is done recursively. It starts from an initial value, and in each time step, the computer calculates a new set of values for the variables at the grid from the former ones. Therefore, truncation errors may evolve in time and the calculation may become unstable.

In general, this strategy of imitating the continuous dynamics of PDEs through a generative mechanism of FDEs makes it possible to treat complex systems, that is, systems in which a theoretical model of the dynamics is known, but the system is intractable for reasons of complexity. There is a wide range of applied problems that meet these requirements: The physical laws are well understood, but the interactions of different processes render the entire system 'complex,' that is, intractable. One cannot hope to achieve a solution of the PDEs with traditional mathematical means.

Nevertheless, both the epistemological and the methodological status of simulations are discussed controversially in the philosophy of science. The common view holds that simulations are more or less calculations that profit from the brute force of the computer. The computer is seen as 'number cruncher.' But there is also a heated discussion about the fundamentally new features that make simulation models and the important new class of models and simulations a new instrument of science.[2] As mentioned above, from the very beginning, simulations were seen as a quasi-empirical, experimental approach. "Broadly speaking, such methods will amount to construction of statistical models of given physical situations and statistical experiments designed to evaluate the behavior of the physical quantities involved" (Ulam 1952: 264).

Whereas Ulam praised this as an inspiring source for mathematics, von Neumann's response was less enthusiastic. He considered the 'experimental' approach to be a kind of trick, not completely appropriate to the mathematical problems of fluid dynamics he was struggling with. But he was pragmatic enough to see that the simulation method might open up a new access. And he suggested meteorology as an ideal case for the application of the FDE simulation strategy.

Skepticism in Meteorology

During the first half of the twentieth century, one rather speculative question in meteorology was which conditions and hypotheses would be sufficient to construct a model of the entire atmosphere that would be able to reproduce its behavior at least in a gross manner. Some achievements of the theory of the general circulation existed, but they pertained to very restricted parts such as to *lateral diffusion* (Rossby in the 1930s), or to the *jet stream* (Palmèn and Riehl in the 1940s). Which kinds of interactions were responsible for the global behavior observed remained simply unknown. The physics of hydrodynamics was well known and commonly accepted, but their nonlinear behavior was completely unknown at that time. Furthermore, it was believed that the simple nonlinear equations could not describe the complex behavior of fluids. The reasons for both irregularities and regularities were seen in the infinite influences coming from the outside world. In short, the hypothetico-deductive method was not applicable, because there was no mathematical instrument available that would allow an investigation of hypotheses or models. Thus it was commonly held "that a consistent theory of the general circulation is out of reach" (Lewis 1998: 42).

Directly after the war, von Neumann set up a working group on meteorology at the Institute for Advanced Studies in Princeton, headed by Jule Charney. The goal was to model the fluid dynamics of the atmosphere and to treat the resulting system of PDEs with the newly developed FDE simulation method. "To von Neumann, meteorology was par excellence the applied branch of mathematics and physics that stood the most to gain from high-speed computation" (Charney, cited acc. to Arakawa 2000: 5).

The design of the computer and that of the problems of meteorology would have to co-evolve, von Neumann suggested. Consequently, and already in 1946, he called a conference of meteorologists "to tell them about the general-purpose electronic computer he was building and to seek their advice and assistance in designing meteorological problems for its use" (Arakawa 2000: 5).

The approach of employing computer simulations on the basis of hydrodynamics – that is, with known theoretical basis but unknown dynamic properties – was to become the starting point for climate research as a modern discipline.

The phenomena of global circulation in the atmosphere show an enormous complexity – different processes interact in a highly nonlinear way. This is the reason why weather forecasts are impossible if one wants to make predictions that go beyond a critical period. Weather is, so to speak, a chaotic system. On the other hand, there are phenomena in the atmosphere's dynamics that are regular for long periods of time. To give an example, the so-called *surface westerlies*, continuously blowing winds north of the equator, have been well-known for centuries and were used when crossing the Atlantic Ocean in sailing ships. This difference – stable global patterns on the one side and unstable chaotic behavior on the other side – represents a major characteristic of complex systems.[3]

A Breakthrough: The 'First Experiment' by Phillips

A path-breaking success changed the skepticism concerning modeling the general circulation, and brought this project right into the center of a new scientific discipline. In 1955, Norman Phillips, working at Princeton's Institute for Advanced Studies, succeeded in his so-called *first experiment* in simulating the dynamics of the atmosphere, that is, in reproducing the patterns of wind and pressure in the entire atmosphere within a computer model (Phillips 1956).[4] The development of a simulation model of the general circulation of the atmosphere was celebrated as a major breakthrough. It surprised the experts, because it had been generally accepted that a theoretical modeling approach concentrating on the hydrodynamic equations would hardly be possible. Namely, it was believed that a model of a complex phenomenon has to be more complex than the system to be analyzed. This first attempt to build a simulation model of the entire atmosphere was considered an 'experiment.' This underlines how uncertain the success of this project was. At the same time, the conception of experiment expresses an important aspect for methodology: In simulations, scientists use their models like an experimental set-up.

The simulation model of the 'first experiment' worked with a very coarse spatial discretization of the atmosphere. In the vertical direction, it exhibited only two layers, and horizontally each grid cell covered more than 200,000 km^2. Phillips had to introduce the physical laws that govern the dynamics of the atmosphere. He used only six basic equations (PDEs), which, since then, have been called the 'primitive equations.' They are generally conceived of as the physical basis of climatology. These equations express well-known physics of hydrodynamics – the surprising thing was that only six PDEs were sufficient to reproduce the complex behavior, and Phillips had the skill and luck to make an adequate choice. This physical basis had to be adapted to the grid. The construction of a discrete model is a typical task of simulation modeling. The global and continuous equations of hydrodynamics had to be reformulated in order to calculate the evolution of the relevant variables in time – pressure, temperature, wind speed – step by step at the grid nodes.

In the first stage of the experiment, the initial state was an atmosphere at rest, with no differences in temperature or pressure, and no flow. In the second stage of the experiment, the dynamics was started, that is, the radiation of the sun and the rotation of the earth were added. The atmosphere settled down in a so-called *steady state* that corresponded to stable flow patterns. The tantalizing question was whether the model would be able to reproduce the global flow patterns of the real atmosphere, for instance, the surface westerlies. The result was positive – everyone was impressed by the degree of correspondence. As mentioned above, the experts were skeptical about the possibility of a global (and not far too complicated) model, but the empirical success was convincing. The decisive criterion for success was the adequate imitation of the phenomena, that is, the flow patterns. Because there was no knowledge about the outcome, Phillips' attempt to use a specific set of equations can be understood as an experiment – an experiment on modeling the equations of motion within a computer.

The continuous primitive equations of the atmosphere were by no means solved (that is, integrated in the strict technical sense) by Phillips' simulation experiment. Instead, the phenomena of the atmosphere were imitated by the generative mechanism of the discrete difference equations. The success of the imitation was judged solely by the correspondence between simulated and observed flow patterns. Hence, the validation of simulation results relies on a quasi-empirical strategy.

The success of the simulation experiment was acknowledged immediately and was judged to constitute a theoretical breakthrough. In the same year, E. Eady, the leading theoretical meteorologist in England, formulated far-sightedly: "Numerical integrations of the kind Dr. Phillips has carried out give us a unique opportunity to study large-scale meteorology as an experimental science" (Eady 1956: 536).[5]

And indeed, this experimental approach via simulations played a major role in shaping the emerging discipline of climate research. A. Arakawa (2000) calls this the "epoch-making first phase" of climate simulation modeling.[6] Experimental access to the climate system is a key for climate science. Another example is a researcher who succeeded recently in showing that the Pacific Ocean can exert a considerable influence on the Gulf Stream in the Atlantic Ocean. He discovered this connection by numerical experiments and described his approach in an interview as follows:

> *Q*: "You feed the model and then you wait and see what happens"?
>
> *A*: "Yes, exactly. That is the case – without simulation I would never been able to obtain this result" (transcript from an interview[7]).

This is not the place to discuss further developments in climate simulation (see, for more details, Küppers and Lenhard 2005). Simulations, of course, spread rapidly to very diverse fields. This development can be summarized by stating that science had acquired a new instrument – the simulation method provided a means of studying complex systems.[8]

HIERARCHICAL INTEGRATION AND THE POLITICAL CONTEXT OF APPLICATION

The Centralized Model of Atmosphere: The Unidirectional Forces of Science and Politics

The 'epoch-making first phase' (Arakawa) assigned a key role to the general circulation models (GCMs). In the 1960s, the next stage began, the 'magnificent second phase' in which climate science evolved as a normal scientific research program, centered around the GCMs and concentrated mainly in a couple of research centers in the United States.[9] Already in 1960 the *Geophysical Fluid Dynamics Laboratory* (*GFDL*), which belongs to a section of the US Department of Commerce, was founded in Princeton to follow up on this approach. This was the first institution with the official task of simulating in climate research. Other typical institutions are the *National Center for Atmospheric Research* at Boulder, Colorado, also founded in 1960, or NASA's *Goddard Institute*. The scientific agenda consisted in refining the GCMs, implementing lattices with higher resolutions, and integrating more subprocesses connected to atmospheric dynamics. In short, the GCMs have been growing more or less continuously for about thirty years.

The GCMs form a class of huge simulation models that run on high-speed supercomputers. This requires a considerable effort in funding, although climate research, having started as a part of meteorology, used to enjoy only limited visibility as a scientific discipline. About twenty years ago, circumstances changed almost completely. The climate system became a subject of hot political debate. The so-called greenhouse effect was discovered and was discussed controversially right from the start. Perspectives on the climate system switched radically. Once seen as a stable system, its potential instabilities and changes now became the topic of discussion and investigation.

The field of climate research became one of the most prominent scientific fields in the media. At the same time, funding rose enormously. Climate research was expected to answer – and was in part defined by that demand – the following questions of utmost public, scientific, and political interest:

- Is there actually a change in the climatic system, or are we observing only random fluctuations; that is, can we *detect* climatic change?
- And if there is a change, are we humans a cause of it; is it an anthopogenic change? That is, can we *attribute* the change to a cause?

Some countries decided that they needed research institutes to tackle these questions. Germany, for instance, founded the *Max Planck Institute for Meteorology*. This institute was built around a GCM (in part imported from the United States, then rapidly developed further) and followed more or less the example of the US institutions mentioned above.[10] GCMs occupied a central role in the scientific enterprise, and this role was assured and fostered by political demands. The goal was to predict the climate system's future state. To be applicable in a political context, such predictions need a high degree of reliability and certainty. The high status of the physical laws that constituted the nucleus of GCMs met this political requirement perfectly.

Besides the scientific efforts, there were remarkable, perhaps unprecedented, global institutional efforts. A joint venture of science and policy was undertaken. The UN and the World Meteorological Organization founded the *Intergovernmental Panel on Climatic Change*, IPCC, a global institution with the official task of delivering an assessment of detection and attribution. Every four to five years, the IPCC publishes an *Assessment Report*, a voluminous compilation of the current state of scientific knowledge. A great number of climate researchers worldwide are involved in this IPCC process. The central tools for analysis and prediction are the GCMs building the backbone of the IPCC's assessment reports. The statements derived from these models serve as a basis for political negotiations and decisions such as the Kyoto protocol.[11]

There is a strong demand for integration for political reasons as well: As is well known, climate change as a political issue instantly attracted opposing parties. Leaving aside considerations about political aims, it is obvious that the *reliability* of knowledge about the climate system became a prominent problem. And that amounts to questions on the validity of the simulation models: Are they really realistic? Are there important subprocesses that have not yet been taken into account? Could these influence predictions of the future development of the climate system?

The policymakers' demand for reliable data on the development of the climate system fostered the efforts to integrate all kinds of effects that were believed to influence the dynamics of the atmosphere. This integration was driven by attempts to make the model more realistic. This was important for climate research as well as climate policy.

Figure 1 should visualize the situation as it is commonly viewed by the community: "basically, it is all physics" (interview), and consequently, the primitive equations of the atmospheric GCM constitute the nucleus – governed by the equations of fluid dynamics. This situation is also reflected in the 'architecture' of the research institutions of climate science. Mostly, they are rooted in physics; for example, the GFDL even bears fluid dynamics in its name. Until recently, the directors were physicists as well.

More and more subprocesses have become attached to the core, that is, are being integrated into the simulation model. Ideally, no essential parts or processes should be

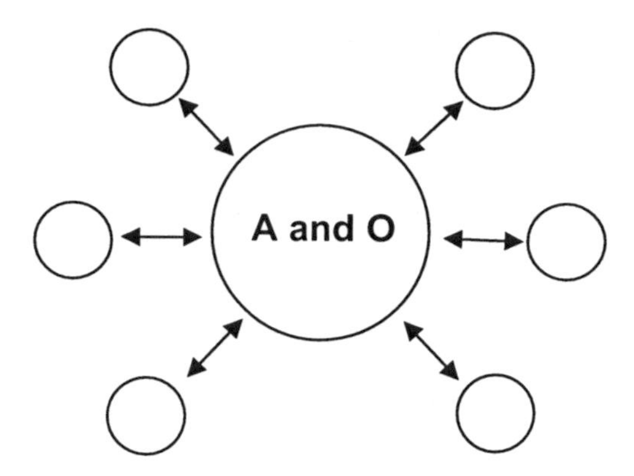

Figure 1. The physics-based atmospheric dynamics at the center. An increasing number of subprocesses are additionally becoming integrated

left out. Integrating aerosols into the GCMs might be seen as a typical success. These gave the models a 'cooling-by-pollution' effect that improved the match with observed temperature patterns.

Far along this line of 'densifying integration,' lies the great achievement of climate modeling in the 1990s: the coupling of atmospheric GCMs and those of oceans (see Figure 2). Both simulation models are centered around hydrodynamic codes – atmosphere and oceans are fluids in physical perspective. This coupling induced no fundamental change in architecture, because physics maintained its position as the theoretical nucleus. The coupled GCMs (CGCMs) once again produce a centralized architecture, now with two centers, resulting in a kind of twin-star image:

A great technical effort was required to couple the two most voluminous simulation models. CGCMs also provided an enriched basis for statistical analyses. The results of CGCM simulations led to a majority opinion that a change of climate can be diagnosed. Moreover, it was a celebrated claim that now, with CGCMs, it became possible to distinguish the so-called 'fingerprint' of human impact.

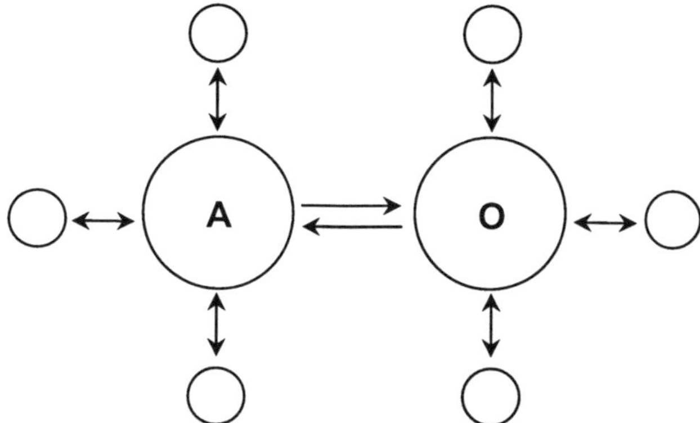

Figure 2. Architecture of coupled atmosphere-ocean generated circulation models (CGCM)

An Episode of Science and Policy: The FA Controversy

While the coupled GCMs were celebrated as a milestone on the road toward a realistic model of the climate system, they gave rise to a heated controversy. The claim to be able to accomplish a more and more comprehensive simulation of the climate system, a simulation drawing on an objective basis of laws of physics, was surely one of the central claims of climate research. For the first time, it became possible to couple atmosphere and oceans. Each system takes the role of a boundary condition for the other. Both systems were calibrated separately to show a steady state similar to the observed phenomena. And now, as the systems became coupled one to the other, the researchers introduced a mechanism to enable interchange while simultaneously guaranteeing that the coupled system would not drift into a new, unrealistic state. In short, this so-called Flux Adjustment (FA) was an 'artificial' mechanism intended to keep the GCMs from leaving their precalibrated (realistic) region.

In some sense, the worst case occurred: In *Science*, one of the most widely read and influential journals, the coupled models were denounced as relying on a "fudge factor" (Kerr 1994). Critics asserted that this flux adjustment was an 'artificial' mechanism without any 'real' counterpart and had been introduced merely to produce the desired results. The coupled models were expected to provide a new and superior integrated basis for predictions, but the criticism of FA challenged this claim. If the results of climate research were not based on 'realistic' models and did not rely on objective laws of physics, would that not question the entire scientific-political enterprise?

The media echo was controversial. The spectrum ranged from 'a blatant scandal' to 'only a storm in a teacup.' Even scientific experts saw things rather differently. We conducted several interviews that also raised this issue. The statements of the scientists ranged from an uncomfortable feeling, because FA was of an artificial nature, across the claim that FA was only a preliminary technique and that the models will be *really* realistic in five years, to the opinion that FA was fully legitimate and comparable to techniques common in simulation modeling.[12]

The heated discussion was accompanied by a critical assessment of the reach of models in general, noteworthy also in *Science* (see Oreskes et al. 1994; see, for a reply defending the modeling approach, Norton and Suppe 2001). None of the sides in this controversy will be taken here. The point is that the incriminated strategy of 'artificial' tools like FA is widely used in simulation modeling and, what is more, belongs to the methodological core of that approach.[13]

Consider, for instance, parameterization in which a complicated mechanism like cloud dynamics is replaced by one or a few parameters that are easier to handle. One could easily extend the criticism against FA to cover parameterization techniques as well – techniques nearly ubiquitous in complex simulation models. Second, the FA affair brings to the fore the hybrid nature of climate research: It is a scientific and political project carried out under the scrutiny of public media. There is a certain tension in the political application of simulation results. Whereas there is no way of treating climatic changes without simulation models, the methodology of simulations seems to cause some tensions with demands for 'realistic' models.

The hybrid scientific-political nature makes it difficult to separate political and scientific motives. The development of 'stars' and ongoing integration up to the twin-star architecture of coupled GCMs can be interpreted in two ways: according to political and according to scientific motives. Hence, the forces of science and policy point in the same direction in this case and result in an ongoing integration. The next section will argue that a fundamental restructuring of the model architecture is presently taking place.

CHANGING THE PARADIGM OF INTEGRATION

From Stars to Networks

The effort to achieve ever greater integration strengthened the star architecture. The emergence of twin stars, that is, coupled atmosphere-ocean models is commonly conceived as a first-rank scientific achievement in the field. Arakawa argues that the 'third phase' of simulation models, that is, integrated modeling, started with this successful coupling.

Up to this point, it is hard to distinguish whether evolution is driven by inner-scientific momentum or induced by political demand. However, the scientific research program of refining and integrating GCMs based on the physics of fluid dynamics has now come to the end of its rope. The paradigm of the centralized model reaches its limiting factors when the processes that are to be integrated have no relation to the theoretical framework. Some leading research institutions are already responding to this by switching to a new architecture of climate simulation models. This new architecture does not deal with integration as an adaptation of additional parts to the dynamic center of GCMs. In fact, one can observe a profound shift in the modeling architecture of simulations in climate science. Roughly speaking, the new approach is to develop models of different, theoretically incompatible fields independently and then to couple them to one another on a merely technical basis of simulation. In this way, it aims at an integration of a variety of models from physics, biology, chemistry, and even economics.[14] Their dynamics can hardly be connected to physics, and therefore the whole architectonic paradigm of a centralized structure seems to be ill-suited for the task of enforced integration. The new architectonic paradigm can be described as a network or grid (see Figure 3).

The most important feature of the new net architecture is that there is no longer one theoretical nucleus. The new nucleus is built by a (virtually theory-free) simulation coupler that is linking the various models. Coupling takes place in a simulation-technical sense (see Winsberg, this volume, who nicely captures the coupling of heterogeneous models as a 'handshake' between them). Each of them has its own theoretical nucleus, thus the net shows symmetry between the models.

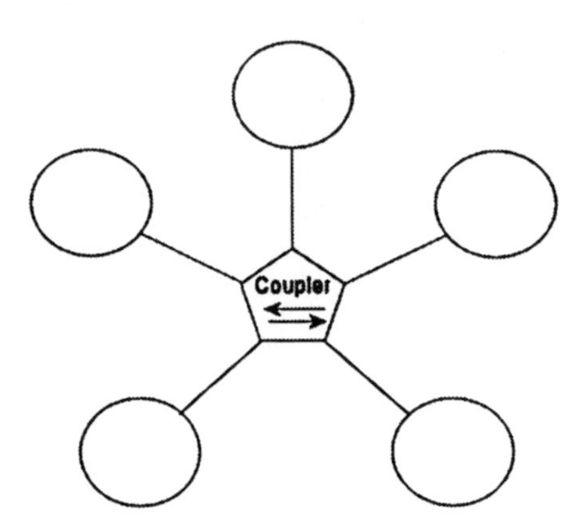

Figure 3. Simulation coupler in the 'void center' integrating various models

The *National Center for Atmospheric Research* (NCAR) has been among the first to implement the new architecture, their program being to realize 'NCAR as an integrator.'[15] The central part of that plan is formed by the so-called *Community Climate System Model* (CCSM) that integrates different simulation models via a hub.

In this organizational structure (Figure 4), a coupler unit controls the exchange of parameter values between independent and exchangeable models. This modeling approach is in clear contrast to earlier attempts at integrating submodels around the center of a physically based GCM.

Thus, the task is no longer to build one all-encompassing model – ideally *the* right model. Instead, researchers construct a model by coupling together different modules that were developed on their own. The coupled network normally presents a

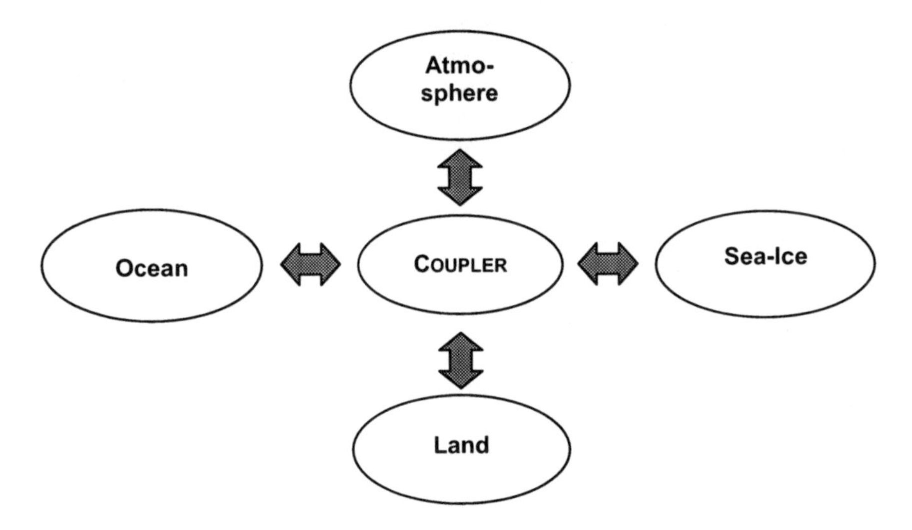

Figure 4. Architecture of NCAR climate simulation modeling

mixture of fully fledged and very basic modules. By replacing some of them, the network can be adjusted to different research questions. The NCAR emphasizes the flexibility of the new net architecture. Figure 4 is explained with the words:

> [...] based on a framework that divides the complete climate system into component models connected by a coupler. Individual components – ocean, atmosphere, land, and sea-ice – can be exchanged for alternate models, thus allowing different configurations appropriate for different applications (CCSM 2004).

The shift from a centralized 'star' to a net architecture is being rated very highly in methodological terms. In interviews, it has been called a 'revolution in modeling style':

> I think actually it's symptomatic worldwide, [...] we had a modeling framework that we have been using for quite some time, but in the last four, five years we are really pretty much throwing it out the window. We have redone our entire modeling framework from scratch, [...] building a proper framework to allow interaction between different physical and biological components, also taking advantage of advances in computer technology to allow the system to be more flexible (transcript from interview).

This amounts to saying goodbye to the fundamental leading role of physics in climate research. In climate change analysis, the physics of fluid dynamics takes at best a position as *primus inter pares*. Under the perspective of ongoing integration, the whole climate system, including all biological, environmental, economic, and other components, is regarded as *one* system. And for this reason, the model architecture can constitute a paradigm for simulations. The analyzed change from hierarchical to network-like integration, which also took place in a social and disciplinary sense, presents a profound paradigm shift – a revolution in modeling style.

CONCLUSION

The foregoing argumentation is oriented toward the case of climate research and the simulation models it employs. It is intended as a contribution to the nuances of the development and the history of the simulation method. Can one draw conclusions beyond that? Are the observations made here also valid in a more general sense? And if so, in what respect? We shall raise three points that conclude our account while also posing new questions:

1. First, the result of the revolution in modeling style can be called pragmatic integration without theoretical background. This allows the integration of theoretically incompatible models. Even models that are distributed, that is, located in different computers can be integrated by this approach. Recently, we can also observe a movement toward so-called 'distributed computing' in somewhat different, although closely related, respects. One important development in this direction is *computational grids*, that is, clusters of computers that are connected but do not execute a central 'program' but contribute their pieces independently. Thus, a huge amount of computational force – but also tons of data that are only available at widely distributed places – can be gathered, and science is making serious efforts to use this kind of resource to manufacture an instrument for the investigation of complex problems that are currently out of reach. Very diverse

projects are conducted 'on the grid' such as climate forecast (connecting more than 25,000 PCs) or pharmaceutical drug development. Those efforts can be summarized under the title 'e-science.' For instance, in 2004, Germany started a research initiative called d-grid to investigate under what conditions a grid architecture can be used effectively. Again, the conditions are of diverse nature: computational, legal, institutional, and many more. As may be obvious, simulation on the grid promises to be one of the major benefits of computational grids. The impact of the new grid architecture has yet to be determined.

2. Second, the simulation architecture indicates a strong application-oriented influence. This results from the need for unifying integration. However, this integration is not achieved in the sense of a unifying theory, but in a pragmatic sense of tinkering the different autonomous models together. "Modeling is a kind of engineering work. We have the components, but they do not fit. And then, we are knocking, or tinkering, them together such that it works" (transcript from interview, see endnote 7).
   This kind of 'tinkering' may be recognized as a quite general feature of science that is under strong pressure from applications, or even dominated by them. Usually, applied problems do not occur at the rare spots that are neatly covered by scientific theories. The lack of a general and common theoretical framework has to be compensated. Whereas hydrodynamic systems are theoretically well understood, when confronted with the real world, they pose complex problems that curtail the range of the theory very strongly. And, moreover, the questions often transcend the theoretical framework – as was observed in the case of the climate system in which science is removing the theoretical nucleus to one of the nodes of a network. This network is connected computationally by a simulation coupler, but theoretically unconnected! There seem to be a plethora of examples in which applied sciences are driven to pragmatic integrations when confronted with the lack of a "rock-bottom of theory" (see Carrier 2004). In the physics of nuclear fusion, for example, the laws have been well known for decades, but the construction of a concrete fusion reactor poses problems that cannot be solved by that theory. Another example, from economics, is so-called innovation networks that should integrate different parts of knowledge on a purely pragmatic level to enable the development of a new product (see, e.g., Pyka and Küppers 2002). Theoretical integration is not the goal. On the contrary, these networks aim at an effective exchange of bits of knowledge, *although* a common theoretical framework does not exist. In sum, tinkering can be considered to be a 'fingerprint' of science dominated by applications.
3. Third, we have argued that a fundamental change, or even a revolution, in modeling style occurred as a change in the *architecture* of simulation models. That architecture may function like a paradigm is due to the fact that simulations are by no means purely theoretical entities. Simulations are different from calculations or algorithms – they work with concrete implementations. Effective implementations limit the possible range of simulations in a pragmatic sense – they have to run on certain machines in reasonable time. Hence, simulations are technologies that have to be investigated on the basis of their application in scientific

practice (see Humphreys 2004 and the introduction to this book).
This aspect characterizes simulations as scientific instruments. Not only do simulations call for appropriate new mathematics to deal with computational issues, but the scientific *instrument* of simulation and *applied problems* (think of climate prediction) are also intimately related and interact. In the case of the climate system investigated here, the 'pressure' for integration originated from the applied context; and, at the same time, the task of integration was embodied in the model's architecture. In sum, simulations are mathematical instruments with a material basis.

Let us conclude with a last consideration: Until now, the guiding line in our argumentation has been computer simulation and especially computer simulation in climate research. However, points of a more general relevance beyond computer simulation are involved, namely, the role of theory and of scientific disciplines and networks. In all cases in which complexity sets limits to analytical solutions, scientific theory is becoming less important and partly replaced by practical ad hoc strategies in knowledge production. Whereas the empirical basis is simply too weak to back such a general claim, something is definitely going on in the relation between science, theory, and applications. One reason for this dynamics is the increasing complexity of science and technology. For example, the idealizations that could still be made in linear regimes are no longer possible in the nonlinear regimes that many questions and problems demand. But this is only one side of the coin. Although theories may not be predictable and even calculable in the strict sense, they play an important role in finding strategies for practical solutions to a broad variety of problems.

The same holds for the organization of knowledge production. The example of computer simulation, especially in the case of climate models, shows a transition from disciplinary organization of knowledge production to a transdisciplinary form of organization. There is no argument that this transition is caused by the simulation as such – it may be due to complexity. The integration of all kinds of competencies, abilities, and knowledge bases in different fields of modern industrial research and development is a very common observation and shows the same network architecture. However, this does not imply the demise of disciplines.

ACKNOWLEDGMENT

We kindly acknowledge the financial support by the Volkswagen Foundation.

Bielefeld University, Bielefeld, Germany

NOTES

[1] Richtmyer and von Neumann 1947, based on contributions by Ulam, and Metropolis and Ulam 1949 count as founding documents of the Monte Carlo method. See also the compilation of Ulam's papers 1990 and the accounts of Galison 1996, 1997, and Fox Keller 2003.

[2] See, for example, Humphreys 1991; Rohrlich 1991; and Fox Keller 2003 who stress the important role of a new kind of experiments. For a more detailed epistemological account of simulation as 'imitation of complex dynamics by a suitable generative mechanism' adhering to the second view and discussing the common view critically, see Küppers and Lenhard 2003 and 2004.

[3] This is what in other contexts is called self-organization; see, for more details, Küppers 2002.

[4] For more details of the experiment, see Lewis 1998; for a broader history of ideas on modeling the general circulation of the atmosphere, see Lorenz 1967.

[5] The use of the word "integration" is an indicator of the strong belief in the calculation paradigm.

[6] For a history of climate research using simulation models, see Edwards 2000.

[7] The interviews were performed within a research project conducted by G. Küppers, J. Lenhard, and H. Lücking. The project (2001 until 2004) addressed the epistemic characterization of simulations, and included interviews with researchers at a couple of climate science centers in Germany and the United States. It was part of the research group *Science in Transition* at the IWT, Bielefeld, funded by the Volkswagen Foundation.

[8] The experimental approach to complex systems of PDEs is only one particular instance. The simulation method has shed its skin several times, see Fox Keller 2003, or Schweber and Waechter 2000.

[9] The hegemonial role of GCMs in climate research is commonly acknowledged. It is discussed critically in Shackley 1998 et al.; see, also, the dispute between Demeritt (2001a, 2001b) and Schneider (2001).

[10] For an 'evolutionary tree' of GCMs, see Edwards 2000.

[11] The IPCC process and its character as a hybrid science-policy enterprise have been analyzed extensively in the literature. It is not possible to give an overview here. The anthology of Miller and Edwards (2001) gives an impression of the science studies approach to climate science and is highly recommended.

[12] A comparison of the impact of this discussion on different modeling centers is given in Krück and Borchers 1999.

[13] In modeling terms, the FA is equivalent to 'Arakawa's trick.' For an epistemological investigation and a characterization of simulations as imitations of complex dynamics using artificial mechanisms, see Winsberg 2003; Küppers and Lenhard 2006. See Petersen 2000 for an emphasis on a simulation-oriented philosophy of climate research.

[14] For an account of the intimate relation between the simulation method and (inter-)disciplinary structure, see Lenhard et al. 2006.

[15] Documented in the web: http://www.ccsm.ucar.edu/models/

REFERENCES

Arakawa, A. (2000). "A Personal Perspective on the Early Years of General Circulation Modeling at UCLA", in D.A. Randall (ed.), *General Circulation Model Development*, San Diego: Academic Press, pp. 1–66.

Carrier, M. (2004). "Knowledge gain and practical use: Models in Pure and Applied Research", in D. Gillies (ed.), *Laws and Models in Science*, London: King's College Publications, pp. 1–17.

CCSM (2004). *Community Climate System Model*, Version 3.0, Coupler Documentation, http://www.ccsm.ucar.edu/models/ccsm3.0/cpl6/index.html (acc. November 2005).

Demeritt, D. (2001a). "The Construction of Global Warming and the Politics of Science", *Annals of the Association of American Geographers*, **91:** 307–337.

Demeritt, D. (2001b). "Science and the understanding of science: A reply to Schneider", *Annals of the Association of American Geographers,* **92:** 345–348.
Eady, E. (1956). "Discussions", Quarterly Journal of the Royal Meteorological Society, **82:** 535–539.
Edwards, P.N. (2000). "A brief history of atmospheric general circulation modeling", in D.A. Randall (ed.), *General Circulation Development, Past Present and Future: The Proceedings of a Symposium in Honor of Akio Arakawa*, New York: Academic Press, pp. 67–90.
Galison, P. (1996). "Computer simulations and the trading zone", in P. Galison and D.J. Stump (eds.), *The Disunity of Science: Boundaries, Contexts, and* Power, Stanford, CA: Stanford University Press, pp. 118–157.
Galison, P. (1997). *Image and Logic: A Material Culture of Microphysics,* Chicago and London: Chicago University Press.
Humphreys, P. (1991). "Computer simulations", in A. Fine, M. Forbes, and L. Wessels (eds.), *PSA 1990,* vol. 2, East Lansing, MI: Philosophy of Science Association, pp. 497–506.
Humphreys, P. (2004). Extending Ourselves. Computational Science, Empiricism, and Scientific Method, New York: Oxford University Press.
Keller, E.F. (2003). "Models, simulation, and 'computer experiments'", in H. Radder (ed.), *The Philosophy of Scientific* Experimentation, Pittsburgh: University of Pittsburgh Press, pp. 198–215.
Kerr, R.A. (1994. "Climate change – Climate modeling's fudge factor comes under fire", *Science,* **265:** 1528.
Krück, C.C. and J. Borchers (1999), "Science in politics: A comparison of climate modelling centres", *Minerva,* **37:** 105–123.
Küppers, G. (2002). "Complexity, self-organisation and innovation networks: A new theoretical approach", in A. Pyka and G. Küppers (eds.), *Innovation Networks, Theory and Practice,* Cheltenham, UK: Edward Elgar Publishing, pp. 22–52.
Küppers, G. and J. Lenhard (2004). "The controversial status of computer simulations", *Proceedings of the 18th European Simulation Multiconference* (2004), pp. 271–275.
Küppers, G. and J. Lenhard (2005). "Computersimulationen: Modellierungen zweiter Ordnung", *Journal for General Philosophy of Science,* **36** (2): 305–329 (to appear 2006).
Lenhard, J., H. Lücking and H. Schwechheimer (2006). "Expertise, mode 2, and scientific disciplines: Two contrasting views", to appear in *Science and Public Policy.*
Lewis, J.M. (1998). "Clarifying the dynamics of the general circulation: Phillips's 1956 experiment", *Bulletin of the American Meteorological Society,* **79:** 39–60.
Lorenz, E. (1967). *The Nature of the Theory of the General Circulation of the Atmosphere*, Geneva: World Meteorological Organization WMO, No. 218, TP. 115: 161.
Metropolis, N. and S. Ulam (1949). "The Monte Carlo Method", *Journal of the American Statistical Association,* **44:** 335–341.
Miller, C.A. and P.N. Edwards (2001). *Changing the Atmosphere,* Cambridge, MA: MIT: Press.
Morrison, M. (1999). "Models as autonomous agents", in M.S. Morgan and M. Morrison (eds.), *Models as Mediators. Perspectives on Natural and Social Science,* Cambridge: Cambridge University Press, pp. 38–65.
Neumann von, J. and R.D. Richtmyer (1947). "Statistical methods in neutron diffusion", in S.M. Ulam, A.R. Bednarek and F. Ulam (eds.), *Analogies Between Analogies. The Mathematical Reports of S.M. Ulam and his Los Alamos Collaborators*, Berkeley and Los Angeles, CA: University of California Press, pp. 17–36.
Norton, S.D. and F. Suppe (2001). "Why atmospheric modeling is good science", in C.A. Miller and P.N. Edwards (eds.), *Changing the Atmosphere*, Cambridge, MA: MIT Press, pp. 67–105.
Oreskes, N., K. Shrader-Frechette and K. Belitz (1994). "Verification, validation and confirmation of numerical models in the earth sciences", *Science,* **263:** 641–646.
Petersen, A.C. (2000). "Philosophy of climate science", *Bulletin of the American Meteorological Society,* **81:** 265–271.
Phillips, N. (1956). "The general circulation of the atmosphere: A numerical experiment", *Quarterly Journal of the Royal Meteorological Society,* **82:** 123–164.
Pyka, A. and G. Küppers (eds.), (2002). *Innovation Networks: Theory and Practice,* Cheltenham, UK: Edward Elgar Publishing.
Rohrlich, F. (1991). "Computer simulation in the physical sciences", in F. Forbes and F. Wessels (eds.), *PSA 1990*, vol. 2, East Lansing, MI: Philosophy of Science Association, pp. 507–518.
Schneider, S.H. (2001). "A constructive deconstruction of deconstructionists: A response to Demeritt", *Annals of the Association of American Geographers,* **92:** 338–344.

Schweber, S. and M. Wächter (2000). 'Complex systems, modelling and simulation", *Studies in the History and Philosophy of Modern Physics* **31**(4): 583–609.

Shackley, S., P. Young, S. Parkinson and B. Wynne (1998). "Uncertainty, complexity and the concepts of good science in climate change modelling: Are GCMs the best tools?", *Climatic Change*, **38:** 159–205.

Ulam, S. (1952). "Random processes and transformations", in *Proceedings of the International Congress of Mathematicians 1950*, vol. 2, Providence, RI: American Mathematical Society, pp. 264–275.

Ulam, S. M., A.R. Bednarek and F. Ulam (eds.), (1990). *Analogies Between Analogies. The Mathematical Reports of S.M. Ulam and his Los Alamos Collaborators*, Berkeley and Los Angeles, CA: University of California.

Winsberg, E. (2003). "Simulated experiments: Methodology for a virtual world", *Philosophy of Science*, **70:** 105–125.

CHAPTER 7

MARCEL BOUMANS

THE DIFFERENCE BETWEEN ANSWERING A 'WHY' QUESTION AND ANSWERING A 'HOW MUCH' QUESTION

> But scientific accuracy requires of us that we should in no wise confuse the simple and homely figure, as it is presented to us by nature, with the gay garment which we use to clothe it. Of our own free will we can make no change whatever in the form of the one, but cut and colour of the other we can choose as we please (Hertz [1893] 1962: 28).

INTRODUCTION

Generally, simulations are carried out to answer specific questions. The assessment of the reliability of an answer depends on the kind of question investigated. The answer to a 'why' question is an explanation (van Fraassen 1988: 138). The premises of an explanation have to include invariant relationships (Woodward 2000), and thus the reliability of such an answer depends on whether the domain of invariance of the relevant relationships covers the domain of the question. The answer to a 'how much' question is a measurement. A measurement is reliable when it is an output of a calibrated measuring instrument. Calibration is defined in metrology as a

> "set of operations that establish, under specified conditions, the relationship between values of quantities indicated by a measuring instrument or measuring system, or values represented by a material measure or a reference material, and the corresponding values realized by standards" (IVM 1993: 48).

The idea of a standard is that it is often based upon naturally occurring phenomena when these possess the required degree of stability. These stable phenomena need not necessarily be invariant relationships but could also be stable facts. An example of this approach to base standards on a set of proper invariants is the way in which the international metric organizations aim to link the base units of the International System of Units (meter, kilogram, second, etc.) to the real world through the fundamental constants of physics (velocity of light, Avogadro constant, etc.) that are supposed to be universal and unchanging.

The simulations discussed in this chapter are research activities carried out with and on mathematical models. Therefore, depending on the kind of questions to be answered, a model should fulfill different requirements. The chapter starts with Hertz

J. Lenhard, G. Küppers, and T. Shinn (eds.), Simulation: Pragmatic Construction of Reality, 107–124.

who distinguished three kinds of requirement: consistency, correctness, and appropriateness. Correctness refers to the requirement that the model should contain equations that are representations of the laws of the phenomena to be investigated. In modern terminology, models that fulfill this requirement of correctness are called white-box models. In economics, this requirement of correctness led to the construction of models that were as large and detailed as possible. Simulations carried out on these white-box models showed that the output displayed similar characteristics to those of the phenomenon being studied, which gave a strong support to this kind of modeling. It will be shown in this chapter, however, that this requirement of correctness is not requisite for answering 'how much' questions. Therefore, the ways models are built to produce reliable results differ for both kinds of question. In economics, the approach to build white-box models is the so-called *Cowles Commission program* that dominated econometrics till the 1980s. Doubts whether the obtained model equations are invariant with respect to policy interventions, worded most expressly by Robert Lucas, led to alternative programs of which Lucas's was most influential. In this so-called general-equilibrium approach, a 'good' model is not necessarily a correct one, but should provide 'imitations' that pass a Turing test. In Kydland and Prescott's implementation of Lucas' program, models came to be considered as measuring instruments, which entails that Turing testing is interpreted as a kind of calibration. It will be shown that calibration is understood to mean that models should be gray boxes; that is, they are constructed and assessed according to a system engineering approach of assembling black boxes whose output displays the desired operating characteristics.

This chapter will show that the economic literature on the use of models for simulation purposes reveals a shift in the importance of the requirement of 'correctness,' namely, a shift from an approach in which correctness is the most dominant requirement to an approach in which this requirement has fully disappeared. However, in this shift from white-box modeling to gray-box modeling, the requirement of appropriateness remained crucial.

HERTZ'S MODEL REQUIREMENTS

The tradition of modeling in empirical economics is rooted in the work of James Clark Maxwell, in particular, his ideas about the use of analogies: "that partial similarity between the laws of one science and those of another which makes each of them illustrate the other" (Maxwell [1855] 1965: 156). In other words, to the extent that two physical systems obey laws with the same mathematical form, the behavior of one system can be understood by studying the behavior of the other, better known, system. Moreover, this can be done without formulating any hypothesis about the real nature of the system under investigation.

Heinrich Hertz recognized the value of the concept of formal analogy in trying to understand the essential features of the natural world. For Hertz, representations of mechanical phenomena could only be understood in the sense of Maxwell's dynamical analogies, which is obvious from his definition of a 'dynamical model' in his *Principles of Mechanics Presented in a New Form*:

> A material system is said to be a dynamical model of a second system when the connections of the first can be expressed by such coordinates as to satisfy the following conditions: (1) That the number of coordinates of the first system is equal to the number of the second. (2) That with a suitable arrangement of the coordinates for both systems the same equations of condition exist. (3) That by this arrangement of the coordinates the expression for the magnitude of a displacement agrees in both systems (Hertz [1899] 1956: 175).

From this definition, Hertz inferred that: "In order to determine beforehand the course of the natural motion of a material system, it is sufficient to have a model of that system. The model may be much simpler than the system whose motion it represents" (p. 176).

While the 'model' was still considered as something material, its relation to the system of inquiry should be the same as the relation of the images (*Bilder*) we make of the system to the system itself; namely, that the consequents of the representation, whether material (model) or immaterial (image), must be the representation of the consequents. However, this relationship between representation and system under investigation would allow for many different representations. Hertz, therefore, formulated three requirements a representation should fulfill. First, a representation should be "(logically) permissible," that is, it should not contradict the principles of logic. Second, permissible representations should be "correct," that is, the essential relations of the representation must not contradict the system relations. Third, of two correct and permissible representations of the same system, one should choose the most "appropriate." A representation is more appropriate when it is more distinct, that is, when it contains more of the essential relations of the system; and when it is simpler, that is, when it contains a smaller number of superfluous or empty relations. Hertz explicitly noted that empty relations cannot be altogether avoided: "They enter into the images because they are simply images, – images produced by our mind and necessarily affected by the characteristics of its mode of portrayal" (Hertz [1899] 1956: 2).

In short, the three requirements that a representation of a phenomenon should fulfill are: (1) logical consistency; (2) 'correctness,' that there is correspondence between the relations of the representation and those of the phenomenon; and (3) 'appropriateness,' that it contains the essential characteristics of the phenomenon (distinctness) as simply as possible. Hertz considered the last requirement as most problematic:

> We cannot decide without ambiguity whether an image is appropriate or not; as to this differences of opinion may arise. One image may be more suitable for one purpose, another for another; only by gradually testing many images can we finally succeed in obtaining the most appropriate (Hertz [1899] 1956: 3).

Appropriateness will appear as the crucial requirement for any satisfactory model building process. Every model is necessarily a simplified picture of a phenomenon under investigation, but this simplification should be such that the picture remains appropriate.

AIM FOR CORRECTNESS

To fulfill the requirement of correctness, one should take care that the model is a representation of the relevant laws. One of the central themes of Trygve Haavelmo's (1944) seminal paper *The Probability Approach in Econometrics* was a discussion of the problem how to uncover these laws outside a laboratory. This problem, labeled by him as the problem of autonomy, was worded as the problem of "judging the degree of persistence over time of relations between economic variables," or, more generally speaking, "whether or not we might hope to find elements of invariance in economic life, upon which to establish permanent 'laws'" (Haavelmo 1944: 13). The problem of autonomy results from the fact that real economic phenomena cannot be "artificially isolated from 'other influences'" (Haavelmo 1944: 14). We have to deal with passive observations, and these are "influenced by a great many factors not accounted for in theory; in other words, the difficulties of fulfilling the condition 'other things being equal'" (Haavelmo 1944: 18).

To explore the problem of autonomy, consider the following more concrete problem. Let y denote an economic variable, the observed values of which may be considered as results of planned economic decisions taken by individuals, firms, and so forth. And let us start from the assumption that the variable y is influenced by a number of causal factors, $x_1, x_2, \ldots$.

> Our hope in economic theory and research is that it may be possible to establish constant and relatively *simple* relations between dependent variables, y (of the type described above), and a relatively *small* number of independent variables, x. In other words, we hope that, for each variable, y, to be 'explained', there is a relatively small number of explaining factors the variations of which are practically decisive in determining the variations of y (Haavelmo 1944: 22–23).

Let the behavior of y be determined by a function F:

$$y = F(x_1, x_2), \ldots \tag{1}$$

Then, the way in which the factors x_i might influence y can be represented by the following equation:

$$\Delta y = \Delta F(x_1, x_2, \ldots) = F_1\,\Delta x_1 + F_2\,\Delta x_2 + \ldots \tag{2}$$

The deltas, Δ, indicate a change in magnitude. The terms F_i indicate how much y will proportionally change due to a change in magnitude of factor x_i. Haavelmo distinguished between two different notions of influence, namely, potential influence and factual influence. When F_i differs significantly from zero, then factor x_i has 'potential influence.' The combination $F_i \cdot \Delta x_i$ indicates the magnitude of the 'factual influence' of a factor x_i upon y.

According to Haavelmo, the distinction between potential and factual influence was fundamental.

> For, if we are trying to explain a certain observable variable, y, by a system of causal factors, there is, in general, no limit to the number of such factors that might have a

> *potential* influence upon y. But Nature may limit the number of factors that have a non-negligible *factual* influence to a relatively small number (Haavelmo 1944: 24).

Thus, the relationship $y = F(x_1, \ldots, x_n)$ explains the actual observed values of y, provided that the factual influence of all the unspecified factors together were very small compared with the factual influence of the specified factors $x_1, \ldots, x_n$. However, "our greatest difficulty in economic research" does not lie in establishing simple relations, but rather in the fact the empirically found relations, derived from observation over certain time intervals, are "still simpler than we expect them to be from theory, so that we are thereby led to *throw away* elements of a theory that would be sufficient to explain apparent 'breaks in structure' later" (Haavelmo 1944: 26). The problem is that it is not possible to identify the reason why the factual influence of a factor, say x_{n+1}, is negligible, that is, $F_{n+1} \cdot \Delta x_{n+1} \approx 0$. We cannot distinguish whether its potential influence is very small, $F_{n+1} \approx 0$, or whether the factual variation of this factor over the period under consideration was too small, $\Delta x_{n+1} \approx 0$. We would simply like to 'throw away' the factors whose influence was not observed because their potential influence was negligible to start with. At the same time, we want to retain factors whose influence was not observed only because they varied so much less that their potential influence was veiled.

Haavelmo's design rules for econometrics were considered to be an alternative to the experimental methods of science (Morgan 1990: 262). However, although researchers at the Cowles Commission[1] adopted Haavelmo's 'blueprint' for econometrics (Morgan 1990: 251), they scrapped the term 'autonomy' because it was believed that the model relations were invariant. The reason for believing this was that Haavelmo had pointed out the possibility that the empirically found relationships may be simpler than theory would suggest. This problem could be avoided by building models that were as comprehensive as possible, based on *a priori* theoretical specifications. The Cowles Commission view was that to understand a particular aspect of economic behavior, it is necessary to have a system of descriptive equations. These equations should contain relevant observable variables, be of a known form (preferably linear), and have estimable coefficients. However, "little attention was given to how to choose the variables and the form of the equations; it was thought that economic theory would provide this information in each case" (Christ 1994: 33).

SIMULATION ON CORRECT MODELS

The Cowles Commission's solution to the problem of autonomy was to build more and more comprehensive models. The idea was to build in as many potential influences as possible. In the 1940s, Lawrence Klein was commissioned to build Cowles Commission type models of the United States. The program's aim was to build increasingly comprehensive models to improve their predictability so that they could be used as reliable instruments for economic policy. Irma and Frank Adelman's (1959) computer simulation of the Klein-Goldberger (1955) model of the United States economy – at that time the most advanced macroeconometric model – showed that this model, when shocked by disturbances, could generate cycles with the same characteristics as those of the United States economy. Indeed, the Klein-Goldberger model cycles were remarkably similar to those described as being characteristic of the United States economy by the National Bureau of Economic Research (*NBER*). From this it was concluded that the Klein-Goldberger model was "not very far wrong" (Adelman and Adelman 1959: 621).

The Klein-Goldberger model consisted of twenty-five difference equations with a corresponding number of endogenous variables; it was nonlinear in character and included lags up to the fifth order. The model had been applied to yearly projections of economic activity in the United States with some success, but its dynamic properties were only analyzed using highly simplified assumptions. The innovative element of the Adelmans' research was that, rather than making simplifying assumptions, the complexity of the Klein-Goldberger model was left intact and the equations were programmed for an IBM 650 and simulated for one hundred annual periods. The reason this work was not done earlier was that until then no technology was available to cope with such a task.

Initially, Irma and Frank Adelman were interested in an endogenous explanation of the persistent business fluctuations so characteristic of Western capitalism. Existing theories led to the idea of either dampened or explosive cycles. To the Adelmans, exogenous shocks or externally imposed constraints seemed "rather artificial," so they looked for a "more satisfactory mechanism for the internal generation of a persistent cyclical process" (Adelman and Adelman 1959: 596). The purpose of their paper was to investigate whether the Klein-Goldberger model was a good candidate. The first step in their research was to run the program in the absence of additional external constraints and shocks. The exogenous variables were extrapolated by fitting a least-squares straight line to the postwar data. The result was that the variables in the Klein-Goldberger model grow almost linearly with time. Thus, the endogenous part of the model did not contain an explanation of the oscillatory process. Two conclusions could be drawn: Either the Klein-Goldberger model is 'fundamentally inadequate,' or, to the extent that the behavior of this system constitutes a valid qualitative approximation to that of a modern capitalist society, the observed solution of the Klein-Goldberger equations implies the need to look elsewhere for the origin of business fluctuations. Under the latter assumptions, cyclical analysis would be limited to an investigation of the reaction of the economic system to various perturbations. And, since the Klein-Goldberger model does present a more or less detailed description of the interactions among the various sectors of the economy, it could be utilized in the examination of the mechanism of response to shocks. Random shocks superimposed on the extrapolated values of the exogenous quantities and random shocks introduced into each nondefinitional model equation induced cycles with three- to four-year periods and amplitudes that were "reasonably realistic."

That the amplitudes and the periods of the oscillations observed in this model were "roughly the same as those which are found in practice" (Adelman and Adelman 1959: 611) was seen by the Adelmans as "merely a necessary condition for an adequate simulation of the cyclical fluctuations of a real industrial economy" (Adelman and Adelman 1959: 611). The question now was whether the shocked model could produce business cycles in the 'technical' sense:

> [I]f a business cycle analyst were asked whether or not the results of a shocked Klein-Goldberger computation could reasonably represent a United States-type economy, how would he respond? To answer these questions we shall apply to the data techniques

> developed by the National Bureau of Economic Research (NBER) for the analysis of business cycles (Adelman and Adelman 1959: 612).

A comparison between the characteristics of the cycles generated by the shocked model and the characteristics summarized in the NBER publications of Burns and Mitchell (1946) and Mitchell (1951) was considered to be "quite a stringent test of the validity of the model" by the Adelmans (Adelman and Adelman 1959: 612).

The striking result was that when random shocks of a "realistic order of magnitude" are superimposed on the Klein-Goldberger model equations, the characteristics of the resulting cyclical fluctuations appeared to be similar to those observed in the United States economy:

> The average duration of a cycle, the mean length of the expansion and contraction phases, and the degree of clustering of individual peaks and troughs around reference dates all agree with the corresponding data for the United States economy. Furthermore, the lead-lag relationships of the endogenous variables included in the model and the indices of conformity of the specific series to the overall business cycle also resemble closely the analogous features of our society (Adelman and Adelman 1959: 629).

The Adelmans concluded that "it is not unreasonable to suggest that the gross characteristics of the interactions among the real variables described in the Klein-Goldberger equations may represent good approximations to the behavioral relationships in a practical economy" (Adelman and Adelman 1959: 620).

Irma Adelman wrote for the *International Encyclopedia of the Social Sciences* the second part of the entry *simulation*, namely, on *Simulation of Economic Processes*.[2] She defined simulation as follows:

> 'Simulation' of an economic system means the performance of experiments upon an analogue of the economic system and the drawing of inferences concerning the properties of the economic system from the behavior of its analogue. The analogue is an idealization of a generally more complex real system, the essential properties of which are retained in the analogue (Adelman 1968: 268).

In other words, a good simulation should be performed on an analogue that serves as an appropriate model. She noted, however, that the connotation of simulation among economists is much more restricted: "The term 'simulation' has been generally reserved for processes using a physical or mathematical analogue and requiring a modern high-speed digital or analogue computer for the execution of the experiments" (Adelman 1968: 268–269).

Adelman observed three major sources for the use of simulation techniques in economics. First, both theory and casual observation suggest that an adequate description of the dynamic behavior of an economy must involve complex patterns of time dependencies, nonlinearities, and intricate interrelationships among the many variables governing the evolution of economic activity through time. Simulation techniques permit the use of more realistic analogues to describe real economic systems.

The second source for the use of simulation arises from the need of social scientists to find morally acceptable and scientifically adequate substitutes for the physical scientist's controlled experiments. To the extent that the analogue used in the simulation represents the relevant properties of the economic system under study, results of experimentation with the analogue can be used instead of those that would have been obtained with analogous experiments with the real economy. Since a simulation study can approximate the economy's behavior and structure quite closely, simulation experiments can, at least in principle, lead to conditional predictions of great operational significance.

Finally, the mathematical flexibility of simulation permits the use of this tool to gain insights into many phenomena whose intrinsic nature is still in no way obvious. It is often possible, for example, to formulate a very detailed quantitative description of a particular process before its essential nature is sufficiently well understood to permit the degree of stylization required for a useful theoretical analysis. Studies on the sensitivity of the results to various changes in assumptions can then be used to disentangle the important from the unimportant features of the problem.

She concluded her entry with the remark that "the usefulness of the technique will depend crucially, however, upon the validity of the representation of the system to be simulated and upon the quality of the compromise between realism and tractability" (Adelman 1968: 273). Hence, in her account of the three sources for simulations in economics, she apparently prefers that the models are correct representations of an economy. But her concluding remark emphasizes that appropriateness, as the "quality of the compromise between realism and tractability," is just as crucial.

ROBERT LUCAS' PROGRAM

To Robert Lucas, the Adelmans' achievement signaled a new standard for what it means to understand business cycles: "One exhibits understanding of business cycles by constructing a *model* in the most literal sense: a fully articulated artificial economy which behaves through time so as to imitate closely the time series behavior of actual economics"(Lucas 1977: 11). To see that the Adelmans' test works as a stringent test, Lucas understood that the facts that are to be reproduced should be a list of characteristics providing as much detail as possible. He found these detailed facts in Friedman and Schwartz's *Monetary History* (1963), Mitchell (1951), and Burns and Mitchell (1946). However, Lucas paraphrased the Adelmans' question above as follows:

> The Adelmans posed, in a precise way, the question of whether an observer armed with the methods of Burns and Mitchell (1946) could distinguish between a collection of economic series generated artificially by a computer programmed to follow the Klein-Goldberger equations and the analogous series generated by an actual economy (Lucas 1977: 11).

This paraphrasing is important because the characteristics test of the Adelmans is thereby reinterpreted as a Turing test. This test was originally described by Turing (1950) as an 'imitation game' to investigate the question "Can machines think?" Today, a Turing test is generally described as follows: Reports based on output of the quantitative model and on measurements of the real system are presented to a team of experts. When they are not able to distinguish between the model output and the system output, the model is said to be valid (see, e.g., van Daalen, Thissen, and Verbraeck 1999). As one can see, the test is in principle the same as the Adelmans' test: An observer (interrogator or expert) has to decide whether a distinction can be made between a computer output and output from the 'real' world.

The enormous advantage of Turing's approach to artificial intelligence is that it freed scientists from building replicas of the human mind to achieve machine thinking that meets the standard of human intelligence. In the same way, the characteristics test of the Adelmans freed macroeconometricians from having to build "detailed, quantitatively accurate replicas of the actual economy" (Lucas 1977: 12). Turing testing legitimized Lucas to work with very simple (and therefore unrealistic) models. Lucas's approach was not to aim at models as 'accurate descriptive representations of reality':

> [I]nsistence on the 'realism' of an economic model subverts its potential usefulness in thinking about reality. Any model that is well enough articulated to give clear answers to the questions we put to it will necessarily be artificial, abstract, patently 'unreal' (Lucas 1980: 696).

Lucas argues for the use of 'as-if *p*' assumptions, where *p* is an analogue system, with the same 'superficial' features as the system under study. Lucas said,

> I think it is exactly this superficiality that gives economics much of the power that it has: its ability to predict human behavior without knowing very much about the make up and lives of the people whose behavior we are trying to understand (Lucas 1987: 241).

According to Lucas, the model assumptions need not be assertions about the world:

> A 'theory' is not a collection of assertions about the behavior of the actual economy but rather an explicit set of instructions for building a parallel or analogue system – a mechanical, imitation economy. A 'good' model, from this point of view, will not be exactly more 'real' than a poor one, but will provide better imitations. Of course, what one means by a 'better imitation' will depend on the particular questions to which one wishes answers (Lucas 1980: 697).

In the 'equilibrium business cycle program' dominated by Lucas's instructions, it became standard practice to run an experiment with an artificial economy:

> One of the functions of theoretical economics is to provide fully articulated, artificial economic systems that can serve as laboratories in which policies that would be prohibitively expensive to experiment with in actual economies can be tested out at much lower cost (Lucas 1980: 696).

For his views on 'superficiality', Lucas, on several occasions, acknowledges the influence of Herbert Simon's 1969 publication *The Sciences of the Artificial* (Lucas 1980: 697; 1987: 241; see also Klamer 1984: 47). In following up this reference, Hoover (1995a) shows that Simon (1969) provided the materials that could be used to construct a methodological foundation for calibration (the choice of the model parameters to guarantee that the model precisely mimics some characteristics; see Hoover 1995a: 25).

The central object of Simon's account is an artifact, which he defines as:

> a meeting point – an 'interface' in today's terms – between an 'inner' environment, the substance and organization of the artifact itself, and an 'outer' environment, the surroundings in which it operates. If the inner environment is appropriate to the outer environment, or vice versa, the artifact will serve its intended purpose (Simon 1969: 7).

The advantage of factoring an artificial system into goals, outer environment, and inner environment is "that we can often predict behavior from knowledge of the system's goals and its outer environment, with only minimal assumptions about the inner environment" (Simon 1969: 8). It appears that different inner environments accomplish identical goals in similar outer environments, such as weight-driven clocks and spring-driven clocks. A second advantage is that, in many cases, whether a particular system will achieve a particular goal depends on only a few characteristics of the outer environment, and not on the detail of that environment. So, we "might look toward a science of the artificial that would depend on the relative simplicity of the interface as its primary source of abstraction and generality" (Simon 1969: 9). Thus, as Hoover rightly observes: "Simon's views reinforce Lucas's discussion of models. A model is useful only if it foregoes descriptive realism and selects limited features of reality to reproduce" (Hoover 1995a: 35).

However, this does not mean that we can take any inner environment as long as the model succeeds in reproducing the selected features. In Lucas's view, the ability of models to imitate actual behavior in the way tested by the Adelmans is a necessary, but not sufficient, condition to use these kinds of macroeconometric models for policy evaluation. Policy evaluation requires "invariance of the structure of the model under policy variations" (Lucas 1977: 12). The underlying idea, known as the *Lucas Critique*, is that estimated parameters that were previously regarded as 'structural' in econometric analyses of economic policy actually depend on the economic policy pursued during the estimation period. Hence, the parameters may change with shifts in the policy regime (Lucas 1976). Therefore, the inner environment is only relatively independent from the outer environment:

> The independence of the inner and outer environments is not something which is true of arbitrary models; rather it must be built into models. While it may be enough in hostile environments for models to reproduce key features of the outer environment 'as if' reality was described by their inner environments, it is not enough if they can do this only in benign environments. [...] Simon's notion of the artifacts helps Lucas's both rejecting realism in the sense of full articulation and at the same time, insisting that only through carefully constructing the model from invariants – tastes and technology, in Lucas's usual phrase – can the model secure the benefits of a useful abstraction and generality (Hoover 1995a: 36).

Lucas's 1976 paper is perhaps the most influential and most cited paper in macroeconomics (Hoover 1995b), and it contributed to the decline in popularity of the Cowles Commission approach. The *Lucas Critique* was an implicit call for a new research program. This alternative to the Cowles Commission program involved formulating and estimating macroeconometric models with parameters that are invariant under policy variations and can thus be used to evaluate alternative policies. The only parameters Lucas "hopes" to be invariant under policy changes are the parameters describing "tastes and technology" (Lucas 1977: 12; 1981: 11–12).

CALIBRATED MODELS

Kydland and Prescott's (1982) paradigmatic new-classical equilibrium, real-business-cycle paper is generally acknowledged as the first application of calibration in economics. Kydland and Prescott (1982) introduced calibration to macroeconomics as a means of reducing "dramatically" the number of free parameters of their business-cycle model (Kydland and Prescott 1982: 1361). In a special symposium: 'Computational Experiments in Macroeconomics' in the *Journal of Economic*

Perspectives (1996), Kydland and Prescott explicated the 'tool' they used in their (1982) *Time to Build* paper. Their 'experiment' was an implementation of Lucas' 'equilibrium business-cycle program' by running a simulation experiment with an artificial economy.

According to Kydland and Prescott (1991: 169; 1996: 70–75), any economic computational experiment involves five major steps: (1) *Pose a Question*: The purpose of a computational experiment is to derive a quantitative answer to some well-posed question. (2) *Use Well-Tested Theory*: Needed is a theory that has been tested through use and found to provide reliable answers to a class of questions. A theory is not a set of assertions about the actual economy, rather, following Lucas (1980), defined as an explicit set of instructions for building a mechanical imitation system to answer a question. (3) *Construct a Model Economy*: An abstraction can be judged only relative to some given question. The features of a given model may be appropriate for some question (or class of questions) but not for others. (4) *Calibrate the Model Economy*: In a sense, model economies, like thermometers, are measuring devices. Generally, some economic questions have known answers, and the model should give an approximately correct answer to them if we are to have any confidence in the answer given to the question with unknown answer. Thus, data are used to calibrate the model economy so that it mimics the world as closely as possible along a limited but clearly specified number of dimensions. (5) *Run the Experiment.*

Kydland and Prescott's specific kind of assessment is similar to Lucas' idea of testing, although he did not call it calibration. It was argued above that Lucas' idea of testing is similar to a Turing test. To have the confidence that a computer is intelligent, it should give known answers to familiar questions. To test models as "useful imitations of reality," we should subject them to shocks "for which we are fairly certain how actual economies, or parts of economies, would react. The more dimensions on which the model mimics the answer actual economies give to simple questions, the more we trust its answer to harder questions" (Lucas 1980: 696–697). This kind of testing is similar to calibration as defined by Franklin (1997: 31): "the use of a surrogate signal to standardize an instrument. If an apparatus reproduces known phenomena, then we legitimately strengthen our belief that the apparatus is working properly and that the experimental results produced with that apparatus are reliable."

The 'harder question' Kydland and Prescott wanted their model to answer was "What is the quantitative nature of fluctuations induced by technology shocks?" (Kydland and Prescott 1996: 71). And the answer to this question was that "the model economy displays business cycle fluctuations 70 percent as large as did the U.S. economy" (Kydland and Prescott 1996: 74). In other words, the answer is supposed to be a measurement result carried out with a calibrated instrument.

But what are the economic questions for which we have known answers? Or, what are the standard facts with which the model is calibrated? The answer is given most explicitly by Cooley and Prescott (1995). They describe calibration as a selection of the parameter values for the model economy so that it mimics the actual economy on dimensions associated with long-term growth by setting these values equal to certain "more or less constant" ratios. These ratios were the so-called "stylized facts" of economic growth, "striking empirical regularities both over time and across countries," the "benchmarks of the theory of economic growth" (Cooley and Prescott 1995: 3):

1. Real output grows at a more or less constant rate.
2. The stock of real capital grows at a more or less constant rate greater than the rate of growth of the labor input.
3. The growth rates of real output and the stock of capital tend to be about the same.
4. The rate of profit on capital has a horizontal trend.
5. The rate of growth of output per capita varies greatly from one country to another.
6. Economies with a high share of profits in income tend to have a high ratio of investment to output. (Cooley and Prescott 1995: 3).

Only the first four 'facts' were used. The last two emphasize the differences between countries or economies and are thus not general enough.

The research explicitly aimed at answering 'how-much' questions and certainly no 'why' questions: "In our business cycle studies, we do not try to fit or explain anything. [...] theory is a set of instructions for constructing a model to measure something" (Kydland and Prescott 1997: 210–211).

The business cycle models are considered as stochastic versions of neoclassical growth theory. The standard framework that has been used to study business cycles is the neoclassical growth model with labor-leisure choice (see King, Plosser, and Rebelo 1988; Cooley and Prescott 1995; Cooley 1997). In this artificial world, each household has an endowment of time, each period of which it must divide between leisure and work. The households in this economy supply capital k_t and labor h_t to firms that have access to a technology described by the function $F(k_t, h_t)$. Aggregate output is determined by the production function

$$y_t = e^{z_t} F(k_t, h_t) \tag{3}$$

where z_t is a random productivity parameter. It is assumed that z_t evolves according to the following process:

$$z_{t+1} = \rho z_t + \varepsilon_t \tag{4}$$

where ε_t is distributed normally, with zero mean and standard deviation σ_ε. This productivity shock is the (only) source of the business cycle. It is assumed that the capital stock depreciates at the rate δ, and that consumers add to the stock of capital by investing some amount of the real output each period. Investment in period t, x_t, produces productive capital in period t+1, so that the accounting relation for the aggregate capital stock is:

$$(1+\gamma)(1+\eta)k_{t+1} = (1-\delta)k_t + x_t \tag{5}$$

where η denotes the rate of population growth and γ denotes the long-term real growth rate, which is assumed to be constant according to the first stylized fact.

Households choose consumption c_t and hours of work h_t at each date to maximize the expected discounted value of utility, given their expectations over future prices:

$$\max_{c,x,h} E\{\sum_{t=o}^{\infty} \beta^t (1+\eta)^t u(c_t, 1-h_t)\} \qquad 0 < \beta < 1 \tag{6}$$

subject to sequences of budget constraints: $c_t + x_t \leq y_t$.

Taking account of the stylized facts of growth (see above) leads to the following parametric classes. The third and fourth stylized fact, combined with a stylized fact (not listed above) that the ratio of labor income to total income also has a horizontal trend, imply a so-called Cobb-Douglas production function, which has the form:

$$F(k_t, h_t) = k_t^{\theta} h_t^{1-\theta} . \tag{7}$$

Table 1. Parameter values

Technology					Preferences ('Tastes')			
θ	δ	ρ	σ_ε	γ	β	σ	α	η
0.40	0.012	**0.95**	0.007	0.0156	0.987	**1**	0.64	0.012

Certain features of the specification of preferences are also tied to growth facts: Per capita leisure is approximately constant and real wages have increased steadily. Taken together, these two growth facts imply that the elasticity of substitution between consumption and leisure (1 - h) should be near unity, which implies a Cobb-Douglas kind of function for the composite commodity: $c^{1-\alpha}(1 - h)^{\alpha}$, where α is the share parameter for leisure in the composite commodity. A second stylized fact used to arrive at an expression of the utility function is that the elasticity of substitution between the composite commodity of consumption and leisure is constant and equal to $1/\sigma$. As a result, these two stylized facts imply the following parametric class of preferences:

$$u(c_t, 1-h_t) = \frac{(c_t^{1-\alpha}(1-h_t)^{\alpha})^{1-\sigma} - 1}{1-\sigma} . \tag{8}$$

Because the parameter σ is 'difficult to calibrate,' it is assumed to be one, so that the parametric class is further restricted to:

$$u(c_t, 1\text{-}h_t) = (1\text{-}\alpha)\log c_t + \alpha \log (1\text{-}h_t). \tag{9}$$

Then, the parameters are given values based on using all kinds of statistics (e.g., NIPA) and assumptions (values that are based only on assumption are bold faced, see Table 1).

BUILDING OF INCORRECT BUT APPROPRIATE MODELS

Model building in the general-equilibrium program can be characterized by what Orcutt (1960) in one of the first accounts of simulations in economics called the building-block approach.

> Extensive testing of individual pieces must be carried out before the pieces are assembled, and even after they have been assembled, it frequently may be necessary to modify some pieces. Also in finding and in eliminating the errors in a large and complex computer program it is important to be able to do it piece by piece. And even after it is assembled, it frequently may be desirable to alter a particular operating characteristic, or parameter, or the initial composition of components and their status variables. For these reasons it is highly useful to take the individual components of a model as building blocks and construct them and the over-all model so that they are like the fully plugable components of a modern piece of electronic equipment (Orcutt 1960: 901–902).

In current systems engineering, this approach is better known as modular design.

> A module is a self-contained component with a standard interface to their components within a system. Modular design simplifies final assembly because there are fewer modules than subcomponents and because standard interfaces typically are designed for ease of fit. Each module can be tested prior to assembly and, in the field, repairs can be made by replacing defective modules. Custom systems can be realized by different combinations of standard components; existing systems can be upgraded with improved modules; and new systems can be realized by new combinations of existing and improved modules (White 1999: 475).

In the general equilibrium literature (see, e.g., the two standard survey volumes Cooley 1995 and Stokey and Lucas with Prescott 1989), two modules are always part of the models: a Cobb-Douglas production function (Equation [7]) and the utility function defined by Equation (9). They already incorporate a list of required stylized growth facts and need only to be slightly adapted to meet the wishes of the customer.

As a result of this systems-engineering model building approach, the assessment of this kind of model will differ from the way, for example, models are built à la Cowles Commission cookbooks. In systems engineering, model testing is carried out by validation. Validity of a model is seen as 'usefulness with respect to some purpose.' In a paper on model validation in system dynamics, Barlas (1996) notes that for an exploration of the notion validation, it is crucial to make a distinction between white-box models and black-box models. In black-box models, what matters is the output behavior of the model: "The model is assessed to be valid if its output matches the 'real' output within some specified range of accuracy, without any questioning of the validity of the individual relationships that exists in the model" (Barlas 1996: 185). White-box models, in contrast, are statements on how real systems actually operate in some aspects. Generating an accurate output behavior is not sufficient for model validity; the validity of the internal structure of the model is crucial too. A white-box model must not only reproduce the behavior of a real system, "but also *explain* how the behavior is generated" (Barlas 1996: 185–186).

Barlas 1996 discussed three stages of model validation: direct structural tests, structure-oriented behavior tests, and behavior pattern tests. For white-box models, all three stages are equally important; for black-box models, only the last stage matters. In his paper, Barlas emphasizes the special importance of structure-oriented behavior tests:

> These are strong behavior tests that can provide information on potential structure flaws. Since structure-oriented behavior tests combine the strength of structural orientation with the advantage of being quantifiable, they seem to be the most promising direction for research on model validation (Barlas 1996: 184).

It is, however, striking to see that the information provided by these tests does not give any direct access to the structure (Barlas 1996: 191), in contrast to the direct structure tests. The structure-oriented behavior tests listed in Barlas' paper are similar to the characteristics tests as carried out by the Adelmans, and include the Turing test.

Though Barlas emphasizes that structure-oriented behavior tests are designed to evaluate the validity of the model structure, his usage of the notion of structure needs some further qualification. The way in which he describes and discusses these tests shows that his notion of structure is not limited to correct descriptions of real systems; it also includes other kinds of arrangements. Structure-oriented behavior tests are also 'strong' for the validation of modular-designed models, and, for these models, the term structure refers to the way the modules are assembled. These models – in line with the labeling of the other two types of model – could be called gray-box models and should pass the structure-oriented behavior tests and behavior pattern tests.

Structure-oriented behavior tests cannot be used to distinguish between arrangements. To use Simon's example, these tests are not appropriate to distinguish whether the time is indicated by a weight-driven clock or spring-driven clock. King and Plosser (1994) conducted an Adelmans' test on a standard neoclassical real business cycle model, explored in King, Plosser, and Rebelo (1988) and discussed above in the previous section. It appeared that this model passed the Adelmans' test, a result which King and Plosser found "uncomfortable":

> While no one has claimed that the Adelmans' test or the one we have conducted here represent particularly powerful tests of a model, it is somewhat troubling to us that two models as different as the Klein-Goldberger model and a neoclassical real business cycle model are both able to pass this 'test' (King and Plosser 1994: 436).

To resolve this 'tension,' they mentioned two options. The first option is to admit that the NBER stylized facts "have not been complete or thorough enough to help us distinguish among competing hypotheses." If 'hypothesis' means 'structure,' I do agree. However, the second option is rather surprising:

> Macroeconomists must face the possibility that tests of the sort we conduct here are not only not powerful, but much of what we think we know about economic fluctuations as organized by Burns and Mitchell may be an artifact of their procedures (King and Plosser 1994: 437).

Though I can sympathize with their worry that some of the stylized facts might be more stylized than factual, the reason why they arrive at this worry seems (at least to me) unjustified.

CONCLUSIONS

To answer 'why' questions, we need correct, that is, white-box models, but for 'what-is-the-effect-of' or 'how-much' questions, we could use gray-box models or even black-box models. Gray-box models are assemblies of modules; these are black boxes with standard interface. As a result, these different kinds of models differ in the ways they are assessed. Tests are questions regarding a model to which we already know the answers. If the model reproduces these answers, the model can legitimately be considered appropriate. The validation of appropriateness is purpose-related. A corollary of this view is that tests and the type of questions a model is designed to answer should belong to the same category.

Gray-box models are 'validated' by the kinds of test that in the general-equilibrium literature all fall under the general heading of 'calibration,' with this being defined generally enough to cover all tests that Barlas (1996) called "structure-oriented behavior tests." To trust the results of a simulation to 'what-is-the-effect-of' or 'how-much' questions, the models that are run should be calibrated and need not be causally descriptive or, in other words, be correct.

Models that are built primarily for forecasting purposes belong to the category of black-box models. Therefore, it suffices for these models that they are tested only for their predictive power, which is a smaller subset of 'what-is-the-effect-of' or 'how-much' questions.

University of Amsterdam, Amsterdam, The Netherlands

NOTES

[1] The Cowles Commission for Research in Economics was set up in 1932 and funded by Alfred Cowles specifically to undertake econometric research. The journal *Econometrica*, in which Haavelmo's paper appeared, was run from the Commission. The Cowles Commission's econometric approach developed in the 1940s and 1950s became the standard approach now found in econometric textbooks.

[2] The first part, *Individual Behavior* was written by Allen Newell and Herbert A. Simon; and the third part, *Political Processes*, by Charles F. Hermann.

REFERENCES

Adelman, I. (1968). "Simulation: Economic processes", in D.L. Sills (ed.), *International Encyclopedia of the Social Sciences*, vol. 14, New York: Macmillan and The Free Press.

Adelman, I. and F.L. Adelman (1959). "The dynamic properties of the Klein-Goldberger model", *Econometrica*, **27**: 596–625.

Barlas, Y. (1996). "Formal aspects of model validity and validation in system dynamics", *System Dynamics Review*, **12.3**: 183–210.

Burns, A.F. and W.C. Mitchell (1946). *Measuring Business Cycles*, New York: National Bureau of Economic Research.

Christ, C.F. (1994). "The Cowles Commission's contributions to econometrics at Chicago, 1939–1955", *Journal of Economic Literature* **32**: 30–59.
Cooley, T.F. (ed.), (1995). *Frontiers of Business Cycle Research*, Princeton, NJ: Princeton University Press.
Cooley, T.F. (1997). "Calibrated models", *Oxford Review of Economic Policy*, **13.3**: 55–69.
Cooley, T.F. and E.C. Prescott (1995). "Economic growth and business cycles", in T.F. Cooley (ed.), *Frontiers of Business Cycle Research*, Princeton, NJ: Princeton University Press. pp. 1–38.
Van Daalen, C.E., W.A.H. Thissen, and A. Verbraeck (1999). "Methods for the modeling and analysis of alternatives", in A.P Sage and W.B. Rouse (eds.), *Handbook of Systems Engineering and Management*, New York et al.: Wiley, pp. 1037–1076.
Van Fraassen, B.C. (1988). "The pragmatic theory of explanation", in J.C. Pitt (ed.), *Theories of Explanation*, Oxford, UK: Oxford University Press, pp. 136–155.
Franklin, A. (1997). "Calibration", *Perspectives on Science*, **5**: 31–80.
Friedman, M. and A.J. Schwartz (1963). *A Monetary History of the United States, 1867-1960*, New York: Princeton University Press.
Haavelmo, T. (1944). "The probability approach in econometrics", supplement to *Econometrica*, **12**.
Hertz, H. ([1899] 1956). The Principles of Mechanics Presented in a New Form, New York: Dover.
Hertz, H. ([1893] 1962). *Electric Waves*, New York: Dover.
Hoover, K.D. (1995a). "Facts and artifacts: Calibration and the empirical assessment of real-business-cycle models", *Oxford Economic Papers*, **47**: 24–44.
Hoover, K.D. (1995b). "The problem of macroeconometrics", in K.D. Hoover (ed.), *Macroeconometrics, Developments, Tensions and Prospects*, Boston: Kluwer, pp.1–12.
IVM (1993). *International Vocabulary of Basic and General Terms in Metrology*, 2nd ed., Geneva, Switzerland: International Organization for Standardization.
King, R.G. and C.I. Plosser (1994). "Real business cycles and the test of the Adelmans", *Journal of Monetary Economics*, **33**: 405–438.
King, R.G., C.I. Plosser, and S.T. Rebelo (1988). "Production, growth and business cycles, I. The basic neoclassical model", *Journal of Monetary Economics*, **21**: 195–232.
Klamer, A. (1984). The New Classical Macroeconomics. Conservations with the new classical economists and their opponents, Brighton, Sussex: Harvester Press.
Klein, L.R. and A.S. Goldberger (1955). *An Econometric Model of the United States, 1929-1952*, Amsterdam: North-Holland.
Kydland, F.E. and E.C. Prescott (1982). "Time to build and aggregate fluctuations", *Econometrica*, **50**: 1345–1370.
Kydland, F.E. and E.C. Prescott (1991). "The econometrics of the general equilibrium approach to business cycles", *Scandinavian Journal of Economics*, **93**: 161–178.
Kydland, F.E. and E.C. Prescott (1996). "The computational experiment: An econometric tool", *Journal of Economic Perspectives*, **10.1**: 69–85.
Kydland, F.E. and E.C. Prescott (1997). "A response from Finn E. Kydland and Edward C. Prescott", *Journal of Economic Perspectives*, **11.1**: 210–211.
Lucas, R.E. (1976). "Econometric policy evaluation: A critique", in K. Brunner and A.H. Meltzer (eds.), *The Phillips Curve and Labor Markets*, Amsterdam: North-Holland, pp. 19–46.
Lucas, R.E. (1977). "Understanding business cycles", in K. Brunner and A.H. Meltzer (eds.), *Stabilization of the Domestic and International Economy*, Amsterdam: North-Holland, pp. 7–29.
Lucas, R.E. (1980). "Methods and problems in business cycle theory", *Journal of Money, Credit, and Banking*, **12**: 696–715.
Lucas, R.E. (1987). "Adaptive behavior and economic theory", in R.M. Hogarth and M.W. Reder (eds.), *Rational Choice: The contrast between economics and psychology*, Chicago, IL: The University of Chicago Press, pp. 217–242.
Maxwell, J.C. ([1855] 1965). "On Faraday's lines of force", in W.D. Niven (ed.), *The Scientific Papers of James Clerk Maxwell*, vol. I, New York: Dover, pp. 155–229.
Mitchell, W.C. (1951). *What Happens During Business Cycles*, New York: National Bureau of Economic Research.
Morgan, M.S. (1990). *The History of Econometric Ideas*, Cambridge, UK: Cambridge University Press.
Orcutt, G.H. (1960). "Simulation of economic systems", *The American Economic Review*, **50.5**: 893–907.
Simon, H.A. (1969). *The Sciences of the Artificial*, Cambridge, MA: MIT Press.
Stokey, N.L. and R.E. Lucas, with E.C. Prescott (1989). *Recursive Methods in Economic Dynamics*, Cambridge, MA, and London: Harvard University Press.
Turing, A.M. (1950). "Computing machinery and intelligence", *Mind*, **59**: 433–460.

White, K.P. (1999). "System design", in A.P Sage and W.B. Rouse (eds.), *Handbook of Systems Engineering and Management*, New York et al.: Wiley, pp. 455–481.

Woodward, J. (2000). "Explanation and invariance in the special sciences", *The British Journal for the Philosophy of Science*, **51**: 197–254.

CHAPTER 8

ERIKA MATTILA

STRUGGLE BETWEEN SPECIFICITY AND GENERALITY: HOW DO INFECTIOUS DISEASE MODELS BECOME A SIMULATION PLATFORM?

INTRODUCTION

Promoting and protecting public health is the major task of the National Public Health Institute (*KTL*) in Finland. Its primary goal is to control epidemic infectious diseases nationally. The best known methods that impact on us all are the immunizations we receive since early childhood. Yet, how can the effects of these interventions[1] be followed and examined on a population level? Clinical studies may not be appropriate for assessing the effectiveness of medical interventions or the spread of epidemic infections for ethical, structural, or financial reasons. To overcome these restrictions, a research group of mathematicians, computer scientists, and infectious disease specialists[2] decided to study and develop appropriate models and computer simulations. In this article, based on an empirical case study,[3] I shall show how this group of modelers first built a family of *Haemophilus* models and then, based on the expertise gained during this process, worked hard to turn these models into a simulation-based research tool for more general use in public health work.

During a ten-year-long multidisciplinary modeling project, the initial research goal, to develop a general research tool for the public health administration, could not be addressed directly. Instead, the researchers studied different aspects of infectious disease modeling by building the family of Hib models. To be precise, the modelers focused initially on studying the spread and transmission of *Haemophilus influenzae type b* (Hib) bacteria within, for instance, a family or a daycare group, in other words, in a small, closed population. This helped them to study simple modeling techniques. Later, when a junior epidemiologist joined the research group, the focus shifted toward more sophisticated epidemiological problems such as evaluating the effectiveness of immunizations. *Haemophilus influenzae type b* bacteria can cause life-threatening diseases such as meningitis or septicaemia. These diseases are rare in the Western countries due to the ongoing immunization programs.[4] The research group built altogether ten Hib models, which I call the *family of Hib models*. Each single Hib model was an answer to a specific research question, and researchers applied the

J. Lenhard, G. Küppers, and T. Shinn (eds.), Simulation: Pragmatic Construction of Reality, 125–138.

novel results when tackling new research problems. However, the modeling work was not an example of a process that progressed straightforwardly. The modelers returned to the mathematical and statistical solutions and developed them further to address, for instance, the dynamics of Hib immunity in every detail. In a way, each Hib model functioned as a 'storage space' for the questions studied. Due to this function, I have identified two kinds of element[5] within each single Hib model: those carrying mathematical and statistical solutions and those describing the modeled phenomenon. These elements are not stable, ready-made building blocks; on the contrary, they were 'moulded'[6] and modified in order to address the questions under scrutiny. This practice, studying a clearly specified question within a model by creating mathematical solutions (such as algorithms) to answer it, is called *tailoring* by the modelers themselves. All the members in the family of Hib models shared this trait – they were specifically built to address (i.e., tailored to fit) certain epidemiological questions.

During the final years of the modeling project, the researchers built a population simulation model that differed from the earlier Hib models. The population simulation model consisted of three parts: a demographic model (covering the structure[7] of the Finnish population), a Hib transmission model (including the contact-site structure), and an immunity model (consisting of the immunizations and their effects). Due to its three-part structure, it was possible to follow different factors such as duration of immunity, carriage of infection, and immunization effects on an individual level and relate the findings to the population level. This means that the model enables you to follow, for instance, what happens if a five-year-old boy gets the infection and how likely it is that he will infect his family members and children in his day-care group, and combine these questions with the changes in immunity levels of these populations (so-called herd immunity[8]) revealing possible risks for those who have not been immunized. Whereas these issues had been studied only partially within the family of Hib models, the population simulation model now provided the possibility to follow all these factors simultaneously. The modelers accomplished this by programming a simulation program with a command-line interface, called *Simulator*. The Simulator was used in explaining these processes and calculating predictions. The modelers underlined that within the Simulator, it was possible to explain certain aspects of the phenomenon (such as the proper immunity rate to reach the herd immunity) while simultaneously calculating possible predictive scenarios (such as what happens if a certain proportion of the population refuses to follow the immunization programs). This seems to require both openness toward the structure of the population simulation model and stability and accessibility from the program running the simulations. By following Boumans' typology (this volume) on black-box, white-box, and gray-box models, it appears to me that the model and its simulation program were designed to function simultaneously both as a 'white-box' and a 'black-box' model.[9] The openness and transparency of the model's structure may have prevented its closure, that is, its transformation into a black-boxed, general research tool.

Despite the success of this population simulation model and its simulation program, the modelers failed to attain their initial aim: to produce a general, simulation-based research tool that epidemiologists in public health work could use as a closed or 'black-boxed,' easy-to-use tool. My analysis will focus on this issue: Why does the Simulator fail to function as a general tool, a simulation platform? I call this general research tool, inspired by Keating and Cambrosio's (2000; 2003) concept of a *biomedical platform*, a simulation platform. Keating and Cambrosio consider the biomedical platform as a combination of techniques, reagents, skills, and constituent entities. In a sense, a platform is a context that brings together the modelers' know-how and skills with the research materials such as data, algorithms, and computational capacity. The initial idea was to use the simulation platform to estimate the cost-effectiveness of vaccination schedules in developing countries. To do this, it would have to be possible for the user to easily change the population context and the modeled pathogen[10] on the simulation platform. But this was not the case: The command-line interface required advanced programming skills and the modelers were unable to program a graphical user interface. I shall argue that the step from the specificity to the generality was frustrated.[11] On the one hand, the challenge of organizing interdisciplinary[12] research may serve as an explanation for this. Did the researchers involved in the modeling work manage to cross the disciplinary boundaries? As Star and Griesemer (1989: 387) aptly emphasized, scientific work is heterogeneous and its "central tension is between divergent viewpoints and the simultaneous need for generalizable findings." Could this tension, manifested in the boundary work,[13] have nurtured the disagreements represented in the failure to reach the initial goal? On the other hand, the lack of resources (such as limited computational capacity and expertise in programming) and the uncertainty concerning the usefulness of the simulation platform may have prevented the modelers from taking the final step from the specific, tailor-made models toward the general simulation platform.

The structure of this chapter is as follows. First, I shall analyze the life span[14] of the family of Hib models and examine the specificity of their elements. I shall combine the story of programming the Simulator (and its different versions) with the life span of Hib models to show why it was important to develop the model and its simulation program simultaneously. This also offers us insight into the role of disciplinary tensions in modeling work. Second, I shall study how the elements were moulded, combined in novel ways, or modified to function as parts of the population simulation model. I shall analyze the modeling practices, such as bargaining over the parameter values, in order to describe how the modelers struggled to make the Simulator function properly. Finally, I shall discuss what problems the modelers faced, and why this led them to fail in their efforts to turn the tailor-made models into a simulation platform.

THE MISMATCH BETWEEN THE TAILOR-MADE MODELS AND THE SIMULATOR

The original idea when developing modeling and simulation methods was to overcome some common constraints of epidemiological study. Large-scale follow-up studies are usually restricted for ethical, financial, and structural reasons. Ethical restrictions refer to the research questions that cannot be studied without causing harm to the part of population under scrutiny. Financial and structural constraints are faced in the demanding processes of gathering data. For instance, to take a specimen in order to identify a potential carrier of an infection may be difficult and expensive if serological tests[15] are not available. But these were not the only reasons for developing computer models as part of the 'research equipment' for public health work. The need to construct predictive scenarios, explain how quickly an infection is able to spread within a population, or study 'hypothetical' cases, such as simulating whether a pathogen is likely to become extinct in a population, were a further underlying motivation for making the effort to program the simulation tool. Simulation as a method of study could also help to overcome the scarcity of data available from developing countries. Health organizations, such as WHO, are committed to counseling work to improve health care in those countries. Planning cost-efficient vaccination programs is part of their work. The production of reliable plans requires, for instance, studies on estimating the correct timing of booster vaccinations to ensure that the proposed programs will lead to sufficient immunity in the population.

In order to understand the difficulties in attaining this level of generality, I shall first analyze the simultaneous processes of building the Hib models and programming the simulation tool. This exemplifies the reasons for the *mismatch* between these models and the tool. Second, I shall analyze the modeling work[16] to reveal the tensions appearing on the practical level and hampering the development of the general research tool. Some clarifications of the terms I use are required: The family of Hib models refers to the same set of models I describe as *tailor-made models*. Hib models cover the epidemiological status of the models, and tailor-made addresses them as a specific group of models built in a particular way. The simulation tool is a computer program called a *Simulator*[17] by the researchers.[18] Four versions, which I call A, B,[19] C1, and C2, were programmed during the modeling project. Versions A and C1 were made by affiliated junior programmers who were not full-time researchers in the group. Version C2 was programmed by the mathematician and computer scientist of the group and utilized the code from C1. However, what is the simulation platform? By simulation platform, I refer to the general research tool that was also the main goal of the project. It was meant to facilitate epidemiological research by overcoming the restrictions discussed earlier, but was not fully accomplished. The following summary (Table 1) introduces the different actors involved in the modeling work, their background, and skills.

The tailor-made models are characterized by their structural transparency and openness. These traits remind us of the 'white-box models' introduced by Boumans (this volume) to clarify the processes of model validation. Instead of addressing the validity issue, I shall apply his descriptions of 'white-box,' 'black-box,' and 'gray-box' models. By white-box, he refers to models providing statements on 'how real systems actually operate in some aspects.' In black-box models, only the 'output behavior' produced by the model is important. Gray-box models are modular-design models in which each module is a self-contained component that can be tested separately. I interpret white-box models as representations of the modeled phenomena, whereas black-box models cover the ways in which they produce the output behavior. Gray-box models appear to be in the middle ground, they capture the behavior of the real systems like the white-boxes, but close certain parts into modules like black-boxes. I shall return the these three types of models in the discussion.

Table 1. The groups of actors and their roles in the modeling work, their disciplinary background, and their goals

Role in the division of labor*	Disciplinary and organizational affiliation	Skills and field of expertise	Goals in modeling
Researchers	Senior infectious disease specialists at National Public Health Institute (*KTL*)	Hib epidemiology, vaccine studies, public health work	To develop a tool for clinical use; advanced studies in epidemiology
	PhD student in medicine at *KTL*	Studies in public health and infectious diseases	PhD in public health science, expertise in epidemiology
Modelers	Professor of Biometry at University of Helsinki	Expertise in mathematical modeling, esp. Bayesian probability theory, background in mathematics and statistics	To develop and apply mathematical and statistical modeling methods to health sciences
	PhD students** in Mathematics and Statistics	Master's degree in applied mathematics	PhD in biometry, expertise in mathematical modeling
	Professor of Computer Science at Helsinki University of Technology (*HUT*)	Expertise in artificial life modeling and developing simulation techniques	To apply simulation techniques in a new context
Programmers	Master's students of engineering at *HUT*	Majoring in computer science, specializing in programming	To program the simulation tool

**This table represents researchers, modelers, and programmers as separate groups for the sake of analytical clarity. However, these roles overlapped (the mathematician participated in the programming and modelers certainly were researching the applicability of modeling methods in epidemiological questions).*

***One of them continued his work as a postdoctoral researcher in the project.*

How can we identify transparency and openness in the models? An examination of both the mathematical and statistical solutions and the elements describing the modeled phenomena shows how they were built upon each other. The difficulties faced indicate that the models and their elements may have been structurally misrepresented in the Simulator. As an example of the evolving model's elements, I introduce the first published model, nicknamed the Good-night kiss model. This model studied how *Haemophilus bacteria* were transmitted within a closed population, namely a family. Its nickname originated from the presupposition that the pathogen was transmitted among family members in good-night kisses. The Good-night kiss model was a starting point for developing mathematical, especially probabilistic modeling methods. These methods,[20] which took into account the typical traits of the modeled phenomena such as bacterial behavior, functioned as an element in the model and helped modelers to treat the problem of 'missing data.' Because the data used in the models were gathered for different purposes in the 1980s, certain parameters needed in the modeling were not available. The Good-night kiss model obliged the researchers to familiarize themselves with the specific traits of pathogen transmission: how the individual's status changed between susceptible, carrier, and infected. These changes had been represented in epidemiological studies by an SIS model,[21] which is a simple pattern studied and developed as one element in the Good-night kiss model. Hence, the Good-night kiss model was a single transmission model for examining how the bacteria were transmitted in a small population. It provided the elementary structure similar to all tailor-made models consisting of elements carrying mathematical and statistical solutions and describing the structure and behavior of the phenomena.[22]

While working on that model, the first version (A) of the Simulator was programmed with a graphical user interface. Visuality was an important component in artificial life modeling, and this was part of the expertise brought into the project by the computer scientist. The programming of version A happened on the threshold of the era of personal computers with user-friendly interfaces (running on Microsoft Windows operating system), which explains why the graphical user interface was actually an achievement. Another reason for user-friendliness was that the Simulator was intended for use outside the research group by professionals with varying computational skills. Therefore, these factors promoted its usability, but did not support its functionality. By usability, I refer to the easy access to the program despite the level of programming skills (or lack of them). Functionality means that the intended functions are well-represented in the Simulator. These functions should, for instance, mimic the structure of the models. In other words, the Simulator is not a fully independent, separate program; it has or should have particular ties with the models studied.

When designing the Simulator (A), the main emphasis was to reach some stage of generality – in a way, to 'go beyond the model.' However, the way in which this should have been achieved was not specified, nor was the relation to the Good-night kiss model discussed. Only the usability was emphasized, because the Simulator was intended to become 'a tool for clinical use' outside the original research group. This meant that the program had to be as easy-to-use as possible, so that, for instance, if an epidemic outbreak is located in a day-care group or a military garrison, the general practitioners providing the health care could run the simulations and estimate the number of infected and the spread of infection. Despite its 'easy-to-use' graphical interface, the Simulator was not used. These broader applications could not be performed with the program, because it did not precisely represent even the simple Good-night kiss model (or any other epidemiological model). The program was merely an 'empty box,' an exercise in simulation method and not a well-functioning research tool. The main importance of programming it along with the modeling work was to show that programming such a tool was possible, that it could apply the graphical user environment to reach the user-friendly interface, and, most importantly, that the modelers were not forced to use any of the available ready-made computer-based program packages.[23] Even though version A failed to function as intended, it did have symbolic value in strengthening the simulation approach among the researchers. The initiative to develop a tool for clinical use had its 'protorepresentation' in this exercise. Unfortunately, the modelers could not rely on the simulated results, because of its failure to properly represent an infectious disease model.

The modeling work continued, and the probabilistic models developed were extended to resolve more complicated research problems in epidemiology. The population structure and transmission mechanism studied in the Good-night kiss model were expanded to cover the actual size of the population and represent its age distribution. Whereas the transmission in the early models considered only Hib, different pathogens, so-called cross-reactive bacteria,[24] were incorporated into the models in the later phase. The SIS structure as a simple epidemiological model was extended to cover immunity level and effectiveness of Haemophilus vaccines. This extension also examined the correct timing and amount of booster vaccinations. Extending the elements reveals their mutual interaction. The probabilistic solutions were required in order to reach the detailed level of epidemiological questions. This extension of the models did not yet result in the programming of a well-functioning simulator. 'Moulding' the elements in the course of modeling to function as desired was supported by the transparent and open structure – that the models were white-box ones. In other words, the modelers were able to return to the solutions 'stored' in the elements of the models and tailor them to answer new questions.

However, the openness and transparency of the tailor-made models did not extend to the Simulator. To address this difficulty, I shall now discuss reasons leading to the mismatch. The first version, A, remained an 'empty box' without serious applications. The programming of the third version, C1, took place after knowledge on the dynamics of immunity, spread of infection, contact structure of the population, and vaccination effects had been accumulated by building the tailor-made models. This version was designed to mimic these pieces of information without restrictions. It was programmed in C++ language with a Finnish commentary and run on a UNIX-based operation system. The programming task was delegated to three engineering students. The program was designed to incorporate the different elements into its structure so that new information could be generated by running the simulations. The idea was, once again, to reach the level of generality in order to make extensive predictions on the population under scrutiny. The usability was disregarded in the programming work, and the Simulator had a command-line interface. This was necessary, because the graphical user interface would have been far too much work for the junior programmers within the given time frame. Yet, this decision also limited the usability of the simulation tool within the research group as well. Only those who had been programming it were able to use it. The epidemiologists, who did not have advanced skills in programming, became dependent on the programmers and their schedules when running the simulations to obtain results.

Although this version of the Simulator seemed to 'fit' the models, this did not happen. Because the programmers were not familiar with the previously built models, they did not share the accumulated know-how 'stored' in the models.[25] This resulted in a discrepancy between the tailor-made models and the Simulator, which only mimicked the structure of the models without a sense of how it should function as a general simulation tool based on them. This discrepancy can be reflected in terms of a division of labor: The Simulators A and C1 were programmed by affiliated, but not full members of the research group. The programmers had followed the modeling work for only a limited time period, and they did not have background education in epidemiology or related subjects. They had a focused task: to produce the simulation program, whereas the researchers were committed to the long-term modeling studies, which included cross-disciplinary studies on modeling methods and epidemiology. It seems that one reason for the mismatching lies in this division of labor. The Simulator, in order to function as originally intended, would have required rather extensive knowledge of the models built. This was not attained in such a short-term programming task. Separating the programming from the modeling activities may have caused the failure. Another reason for this lies in the nature of the Hib models.

As I have shown, these models were indeed tailored. The questions addressed were specified and their scope was limited. This specificity was well-reflected in the problems of programming the Simulator. The programmers were incapable of separating the elements of the tailor-made models and upgrading them in the Simulator to the required level of generality. Some of its features, like the population structure with average divorce rates, provided too detailed information that was not necessary for modeling purposes. However, the elements developed in the models provided generalizable know-how despite their specificity. This know-how and expert skills could be exploited when the modelers began to work on the population simulation model examined in the following.

APPROACHING THE SIMULATION PLATFORM

The mismatch was finally overcome when the researchers built the population simulation model and programmed the fourth version of the Simulator C2 themselves. This population simulation model differed from the previously built models through its three-part structure. The demographic model, the transmission model, and the immunity model were its inseparable parts, built by 'moulding' the elements of the tailor-made models. This complex structure was addressed in the Simulator C2. This program was a revised version of C1, utilizing most of the program code and applying the same command-line interface. It was debugged and reprogrammed to mimic all three parts of the population simulation model. The very same goal of achieving a general research tool reemerged, although the graphical interface was left aside. The idea was to reach the level of generality by extending the elements of the tailor-made models, combining them into the three-part structure of the population simulation model, and translating the model into the simulation program, thus upgrading the modeling work into the world of simulations. This was motivated by a desire to study 'what if'-type questions, because answering these questions by making inferences directly from existing data usually means that the possible harm has already happened on the population level. These future-oriented questions and predictive scenarios concerning the correct timing of booster vaccinations, or the mechanisms and changes in herd immunity, were proposed because the very idea of the modeling was to stay 'one step ahead' and to simulate the possible solutions beforehand. In practice, this required reestimating and changing the parameters of the population simulation model in order to obtain decent answers. In other words, this was clearly a description of conducting research on the *simulation platform*. As introduced, the simulation platform, in its comprehensive meaning, is a context that brings together both modeler's know-how and research materials, and these were present in the simulation practices. But this story does not have a 'happy ending.' Next, I shall examine the practice of reestimating the parameters, which will problematize whether the simulation platform was accomplished or not.

What kind of process was the estimation of parameters? The computer, which is dependent on the modelers' input of values, functions by calculating the results and comparing them with the actual values. The idea that the computer tries to use the parameter values that make the model's behavior look very much like your reality that you have observed raises a problem if the reality observed is 'missing.' In other words, the problem emerged, because only some of the parameter values estimated could have been transferred from the previously built models to the population simulation model as such, whereas others were only to be estimated in this model. The estimation was eventually accomplished by searching for the best combination of parameters to achieve the optimal fit with the observed data. This practice did not follow the Bayesian estimation of parameters, which had been applied in the previous tailor-made models. The tension between the different approaches to parameter estimation led the researchers to 'bargain'[26] for the 'correct' parameter values. During the 'bargaining,' the researchers revised the previous models, given that they could have applied the results as parameter values. They also referred to the current literature as a possible source for estimating the correct values. In the end, given these considerations and previously gained know-how and expertise on Hib epidemiology, they agreed on an estimation of correct values.

The estimation of parameters was only one example disclosing the hidden disagreement over the possible modeling method (Bayesian inference or simulation techniques). This disagreement may have functioned as one obstacle to accomplishing the simulation platform. The choice of modeling method was actually a choice between specificity and generality. The Bayesian inference supported the question-specific approach and allowed *tailoring* the models to fit the data by providing tools to tackle their uncertainty. In contrast, the data specificity was, to some extent, indifferent to the simulation techniques, because the general aim was to *generate* new data through simulations.

The question of usability raises another aspect of the difficulty in accomplishing the general research tool. The usability was discarded in the programming of the Simulator C2, because the modelers did not have the time[27] to program the graphical interface or the resources to hire someone to do it for them. Their aim was to use the program initially just in the research group, and perhaps modify it later for broader distribution. Therefore, it still had a command-line interface, and advanced programming skills were required to use it. It was not the easy-to-use, stable platform to be used by modelers, researchers, or epidemiologist outside this research group as originally intended. The instability resulted partly from the re-programming of the previous version. The researchers were unable to fully debug the code due to the limited time resources. They evaluated the results and accepted the 'anomalies' occurring in the simulated population.[28] Its applicability was also restricted to the current (Finnish) population structure and modeled pathogen, Hib. Therefore, the basic functions of a simulation platform, such as to easily define parameter values to represent different pathogens, different populations, and different vaccination programs, were not attained. Some of the researchers argued that the models translated into the Simulator were 'knitted' too tightly with each other and that functioning as a platform would have required a loose, separable structure. Others opposed this view and emphasized that the three-part structure of the model underlying the Simulation platform was loose enough to allow these changes. Clearly, a closure was not achieved.

On the epistemic level, the initial tensions, partially arising from the different approaches to modeling work, prevented the researchers from acknowledging the achievement of the Simulator. Some regarded it as a failure, and thought that the question-specific modeling was, in the end, the more accurate way to examine the epidemiological issues. This was reflected in questions regarding whether the generalizability in the form of simulation platform was in any way desirable. Is tailoring a more appropriate and efficient way to analyze infectious disease dynamics? In contrast, other researchers regarded the simulation tool as the most important step taken and supported the idea that it should be developed further, re-programmed, and stabilized. They saw it as a potential research tool, as a simulation platform (or, at least, its initial stage). The questions addressing the ambiguity between specificity and generality actually reveal the tension between the open and closed structure of the models, the white-box versus the black-box models. It seems to me that the level of generality required in the simulation platform could not have been reached, because the modeling practice was heavily restrained by tailoring the models. While keeping the structure of the models transparent, 'moulding' their elements taught the modelers to examine specific questions and build reliable models to answer them. However, this was not enough to reach the general level of a simulation platform, which would have required closing and 'black-boxing' certain elements in the course of modeling.

CONCLUDING REMARKS

What were the obstacles to reaching the simulation platform? As the analysis shows, the hindrances have a dual origin. First, they lie in the specialized division of labor and the variation in the skills and expertise of the researchers involved in the modeling task. Second, the inappropriate techniques applied when programming the Simulator, and the Simulator's incapability to represent the models adequately prevented the achievement of the general research tool, the simulation platform.

The specialized division of labor was determined by the disciplinary expertise and skills of the researchers. This need not present an unsurmountable problem. As Küppers and Lenhard's case study (this book) on climate research has shown, coupling, or even tinkering, together models from different disciplines is possible. In such a case, however, the architecture deserves high priority right from the beginning. Hence, one can assume that the task of integration has been underestimated in the Hib case. This was apparent when the programming was delegated to junior engineering students as a clearly focused task. They were not capable of crossing the disciplinary boundaries and absorbing the accumulated knowledge gained through the modeling, and this resulted in the mismatch between the models and the simulation tool. Furthermore, divergent, discipline-informed viewpoints were lurking behind the technical decisions on how to fit the models with the simulation tool. This tension was well documented in the bargaining for the correct parameter values, and it reminds us of the invisible boundaries to be found in interdisciplinary research.

Generally speaking, the programming of the simulation platform can be seen as the development of a new architecture. It is a similar process to Küppers and Lenhard's (this volume) description of 'pragmatic integration,' which means integration of theoretically incompatible models in a 'sense of tinkering different autonomous models together.'[29] If regarded as a form of 'pragmatic integration,' the platform architecture functions as a research tool to mimic, for instance, the patterns of bacterial behavior. Its origin in the tailor-made models instead of simulation techniques prevented its transformation into a more general, independently applicable method of computation.

In Boumans' terminology, the tailor-made models provided a niche of white boxes with their open and transparent structure. The platform, in order to function as desired, would have required black-boxing to some extent. This may also be suggested by Winsberg's account (this book) of how models perform a 'handshake,' even though their dynamics are not mutually consistent. In the end, the questions concerning the appropriate approach remain unanswered. Could this inability to reach closure actually imply that neither method was the most suitable one? It seems to me that, in its current form, the platform remains in the middle[30] between the specificity of the tailoring and the intended generality of simulating. To fulfil the requirements of functionality and usability, certain parts of the platform should have been fixed and simplified, perhaps by following the modularity of Boumans's gray-box models. However, by remaining in the middle position, the platform succeeds in offering some promise of a new approach to the study of complex phenomena.

ACKNOWLEDGMENTS

I would like to thank the researchers at the National Public Health Institute, the University of Helsinki, and the Helsinki University of Technology for their support toward my study. I am grateful for the reviewers and editors, especially Dr. Terry Shinn, for their valuable comments and insights on this article. I wish to thank Professor Reijo Miettinen and Professor Mary Morgan for their comments on previous versions. Funding from the project *Changing University Research and Creative Research Environments* (No. 49789) provided by the Academy of Finland, from the Finnish Post-Graduate School for Science and Technology Studies, and *Nature of Evidence: How Well Do 'Facts' Travel*, Leverhulme Trust/Economic and Social Research Council project at Economic History Department, London School of Economics and Political Science is acknowledged. Part of the research was conducted by the author during 2004-2005 when she was a Research Scholar at the Department of Economic History at the London School of Economics and Political Science, funded by a grant from the Academy of Finland and the Helsingin Sanomat Centennial Foundation.

London School of Economics and Political Science, United Kingdom, and University of Helsinki, Finland

NOTES

1 Immunizations are medical interventions.

2 They have a background in medicine with specializations in epidemiology and public health.

3 The data were collected during 2001–2004 in collaboration with a research group studying the mathematical modeling of infectious diseases. The dataset consists of ethnographic observations during the research group meetings, interviews with the key actors, researchers' scientific publications, and archived documents.

4 In Finland, for example, children have been immunized since mid-1980s.

5 By the concept of element, I refer to the building blocks of models. It is close to Boumans' (1999) notion of the *ingredients* of models.

6 Boumans' (1999) terminology.

7 Population structure includes typical family size and the age distribution of the population.

8 Herd immunity means that if the recommended immunizations take place in a population, those who cannot be immunized (due to a chronic illness or pregnancy) are also protected.

9 Boumans (this volume) discusses the relation between the questions studied within different types of the models and the problem of the models' validity. He argues that black-box models, which are usually forecasting models, are evaluated on the basis of their output. White-box models are validated by examining both models' inner structure and output; they are normally used in answering why-questions. 'How much-' and 'what is the effect of-'questions are answered in gray-box models.

10 Both bacteria and viruses are pathogens.

11 The dynamics of specificity and generality in simulations are discussed by Küppers, Lenhard, and Shinn (introduction to this volume) and Johnson (this volume).

12 I have studied the characteristics of interdisciplinary modeling in Mattila (2005).

13 The concept of boundary work (and the broader discussion on boundary crossing, boundary objects, etc.) has a long tradition in the social sciences. Originally, the analysis focusing on the boundaries in science was presented in Gieryn (1983) and Star and Griesemer (1989). Klein (e.g., 1996) has developed the concepts in her study of the organization of interdisciplinary research. By boundary work, Gieryn (1983: 792) refers to the professionalization of science, which has resulted in ideological

demarcations of disciplines, specialities, or theoretical orientations. This is seen in the ambiguous, flexible, historically changing, and sometimes disputed boundaries of science.

[14] Daston (1999) provides an inspiring starting point for analyzing the life span of research objects. van den Bogaard (1999) has applied a similar approach in relation to economic models.

[15] This is the case with Hib. The only way to identify the infection is to take a specimen from the human nasopharynx.

[16] To be understood in a broad sense covering also the programming of the Simulator.

[17] Although it does not function as any kind of training program like a flight simulator.

[18] The research group of five mathematicians, one computer scientist and one epidemiologist (altogether seven researchers) remained almost unchanged throughout the ten-year research project. The minor changes are not relevant in this story.

[19] The version B is of no interest here, because it was programmed by a visiting researcher and not used by the modelers.

[20] The modelers specialized in Bayesian inference, which is a statistical paradigm for estimating unknown (model) parameters in terms of their probability distribution as a function of the observed data (Leino 2003: 7).

[21] S-I-S (Susceptible-Infective-Susceptible) is a model used for describing recurrent infections (Leino 2003:7).

[22] These elements are analyzed in detail in Mattila (2005).

[23] For instance, Guala (2002) has discussed how computer-based modeling packages are used in research.

[24] These are pathogens that influence humans and thus strengthen their immunity indirectly.

[25] The modelers underlined that one important function of the models was that they offered 'storage space' for the know-how gained during modeling.

[26] The 'bargain' took place in email discussions between two meetings in spring 2002.

[27] The time shortage was tied to the practical goal of the project: to finish a PhD study in medicine during the funding period.

[28] One anomaly was called 'Jesus child,' because during the simulation run, one child aged 2000 years was regularly born in the simulated population.

[29] This reminds us also of Winsberg's (this volume) account of multiscale methods used in nanoscale modeling, which aim at coupling together different levels of description in a seamless model by finding the appropriate handshaking algorithms.

[30] The middle position also has implications for the validity of the simulation platform. Openness and transparency of white-box models actually implies that their validity should be examined in terms of both their inner structure and output. In the case of our platform, the validity of the results should be examined in a similar way. This issue calls for a further study.

REFERENCES

Boumans, M. (1999). "Built-in justification", in M. Morgan and M. Morrison (eds.), *Models as Mediators: Perspectives on Natural and Social Science*, Cambridge, UK: Cambridge University Press, pp. 66–96.

Daston, L. (1999). *Biographies of Scientific Objects*, Chicago: University of Chicago Press.

Gieryn, T. (1983). "Boundary-work and the demarcation of science from non-science: Strains and interests in the professional ideologies of scientists", *American Sociological Review*, **48** (6): 781–795.

Guala, F. (2002). "Models, simulations and experiments", in L. Magnani and N. Nersessian (eds.), *Model-Based Reasoning: Science, Technology, Values,* New York: Kluwer, pp. 59–74.

Keating, P. and A. Cambrosio (2000). "Biomedical platforms", *Configurations*, **8**: 337–387.

Keating, P. and A. Cambrosio (2003). *Biomedical Platforms. Realigning the Normal and the Pathological in Late-Twentieth-Century Medicine*, Cambridge, MA: The MIT Press.

Klein Thompson, J. (1996). *Crossing Boundaries, Knowledge, Disciplinarities and Interdisciplinarities*, Charlottesville, VA: University Press of Virginia.

Leino, T. (2003). *Population Immunity to Haemophilus Influenzae Type B – Before and After Conjugate Vaccines,* Publications of the National Public Health Institute A23/2003. Helsinki, Finland: Hakapaino.

Mattila, E. (2005). "Interdisciplinarity 'in the making': Modeling infectious diseases", *Perspectives on Science: Historical, Philosophical and Sociological*, **13** (4): 531–553.

Morgan, M. and M. Morrison (1999). *Models as Mediators: Perspectives on Natural and Social Science,* Cambridge, UK: Cambridge University Press

Star Leigh, S. and J. Griesemer (1989). "Institutional ecology, 'translations' and boundary objects: Amateurs and professionals in Berkeley's Museum of Vertebrate Zoology, 1907–39", *Social Studies of Science*, **19**: 387–420.

van den Bogaard, A. (1999). "Past measurements and future prediction", in M. Morgan and M. Morrison (eds.), *Models as Mediators: Perspectives on Natural and Social Sciences*, Cambridge, UK, Cambridge University Press, pp. 282–325.

CHAPTER 9

ERIC WINSBERG

HANDSHAKING YOUR WAY TO THE TOP: SIMULATION AT THE NANOSCALE*

INTRODUCTION

All the pundits, prognosticators, and policymakers agree: Research into the science and technology of the nanoscale is going to be one of the hot scientific topics of the twenty-first century. According to the web page of the *National Nanotechnology Initiative*, moreover, this should make nanotechnology and nanoscience 'of great interest to philosophers.' Admittedly, the kind of philosophy being imagined by the authors of the web page initiative is most likely something like the nanotechnological analogue of bio-ethics – not the kind of philosophy typically practiced by the current professional community of philosophers of science. But what about us?

Should we philosophers of science, those of us who are interested in methodological, epistemological, and metaphysical issues in the sciences, be paying any attention to developments in nanoscale research? Undoubtedly, it is too early to tell for sure. But arguably, the right *prima facie* intuition to have is that we should. After all, major developments in the history of the philosophy of science have always been driven by major developments in the sciences. It is true that, historically, most of those scientific developments have involved revolutionary changes at the level of fundamental theory (especially, of course, the revolutionary changes in physics at the beginning of the twentieth century.) It is also true that nanoscience is unlikely to bring about innovations in fundamental theory. But, surely, there is no reason to think that new experimental methods, new research technologies, or innovative ways of solving a new set of problems within existing theory could not have a similar impact on philosophy. And it is not altogether unlikely that some of the major accomplishments in the physical sciences to come in the near future will have as much to do with modeling complex phenomena within existing theories as they do with developing novel fundamental theories.

So far, none of this is meant to be an argument, but simply an impressionistically motivated suggestion that nanoscience *might* be something of philosophical interest.

J. Lenhard, G. Küppers, and T. Shinn (eds.), Simulation: Pragmatic Construction of Reality, 139–151.

The project of this chapter is to look and see, and to try to give a more informed answer. Because of my past work, the place that I am inclined to do that looking is in aspects of model-building, especially the methods of computer simulation that are employed in nanoscience. What I find is that it does indeed look as if there are good prospects for philosophers of science to learn novel lessons, especially about the relations between different theories, and between theories and their models, by paying attention to developments in simulation at the nanoscale.

To begin: What exactly is 'nanoscale science?' No precise definition is possible. But, intuitively, it is the study of phenomena and the construction of devices at a novel scale of description: somewhere between the strictly atomic and the macroscopic levels. Theoretical methods in nanoscience, therefore, often have to draw on theoretical resources from more than one level of description.

Take, for example, the field of nanomechanics. 'Nanomechanics' is the study of solid-state materials that are too large to be manageably described at the atomic level and too small to be studied using the laws of continuum mechanics. As it turns out, one of the methods of studying these nanosized samples of solid-state materials is to simulate them (i.e., study them with the tools of computer simulation) using hybrid models constructed out of theories from a variety of levels (Nakano et al. 2001). As such, they create models that bear interestingly novel relationships to their theoretical ancestors. So, a close look at simulation methods in the nanosciences could offer novel insights into the kinds of relationship that exist between different theories (at different levels of description) and between theories and their models.

If we are looking for an example of a simulation model likely to stimulate those sorts of insights, we need look no further than so-called 'parallel multiscale' (or sometimes 'concurrent coupling of length scales,' CLS) methods of simulation. These methods were developed by a group of researchers interested in studying the mechanical properties of intermediate-sized solid-state materials (how they react to stress, strain, and temperature). The particular case that I shall detail below, developed by Farid Abraham and a group of his colleagues, is a pioneering example of this method.[1] What makes the modeling technique 'multiscale' is that it couples together the effects described by three different levels of description: quantum mechanics, molecular dynamics, and continuum mechanics.

MULTISCALE MODELING

Modelers of nanoscale solids need to use these multiscale methods – the coupling together of different levels of description – because each individual theoretical framework is inadequate on its own at the scale in question. The traditional theoretical framework for studying the mechanical behavior of solids is continuum mechanics (CM). CM provides a good description of the mechanics of macroscopic solids close to equilibrium. But the theory breaks down under certain conditions. CM, particularly the flavor of CM that is most computationally tractable – linear elastic theory – is no good when the dynamics of the system are too far from equilibrium. This is because linear elastic theory assumes that materials are homogeneous even at the smallest scales, when, in fact, we know this is far from the truth. It is an idealization. When modeling large samples of material, this idealization works, because the sample is large enough so that one can effectively average over the inhomogeneities. Linear elastic theory is in effect a statistical theory. But as we get below the micron scale, the fine-grained structure begins to matter more. When the solid of interest becomes smaller than approximately one micron in diameter, this 'averaging' fails to be adequate. Small local variations from mean structure, such as material decohesions – an actual tearing of the material – and thermal fluctuations, begin to play a significant role in the system. In sum, CM cannot be the sole theoretical foundation of 'nanomechanics' – it is inadequate for studying solids smaller than one micrometer in size (Rudd and Broughton 2000).

The ideal theoretical framework for studying the dynamics of solids far from equilibrium is classical molecular dynamics (MD). This is the level at which thermal fluctuations and material decohesions are most naturally described. But computational issues constrain MD simulations to about 10^7-10^8 molecules. In linear dimensions, this corresponds to a constraint of only about fifty nanometers.

So MD methods are too computationally expensive, and CM methods are insufficiently accurate, for studying solids that are on the order of one micron in diameter. On the other hand, parts of the solid in which the far-from-equilibrium dynamics take place are usually confined to regions small enough for MD methods. So the idea behind multiscale methods is that a division of labor might be possible – use MD to model the regions where the real action is, and use CM for the surrounding regions where things remain close enough to equilibrium for CM to be effective.

There is a further complication. When cracks propagate through a solid, it involves the breaking of chemical bonds. But the breaking of bonds involves the fundamental electronic structure of atomic interaction. So methods from MD (which use a classical model of the energetic interaction between atoms) are unreliable right near the tip of a propagating crack. Building a good model of bond breaking in crack propagation requires a quantum mechanical (QM) approach. Of course, QM modeling methods are orders of magnitude more computationally expensive than MD. In practice, these modeling methods cannot model more than two hundred and fifty atoms at a time.

The upshot is that it takes three separate theoretical frameworks to model the mechanics of crack propagation in solid structures on the order of one micron in size. Multiscale models couple together the three theories by dividing the material to be simulated into three roughly concentric spatial regions. At the center is a very small region of atoms surrounding a crack tip, modeled by the methods of computational QM. In this region, bonds are broken and distorted as the crack tip propagates through the solid. Surrounding this small region is a larger region of atoms modeled by classical MD. In that region, material dislocations evolve and move, and thermal fluctuations play an important role in the dynamics. The far-from-equilibrium dynamics of the MD region is driven by the energetics of the breaking bonds in the inner region. In the outer region, elastic energy in dissipated smoothly and close to equilibrium on length scales that are well modeled by the linear-elastic, continuum dynamical domain. In turn, it is the stresses and strains applied on the longest scales that drive the propagation of the cracks on the shortest scales (see Figure 1).

It is the interactions between the effects on these different scales that lead students of these phenomena to describe them as "*inherently* multiscale" (Broughton et al.

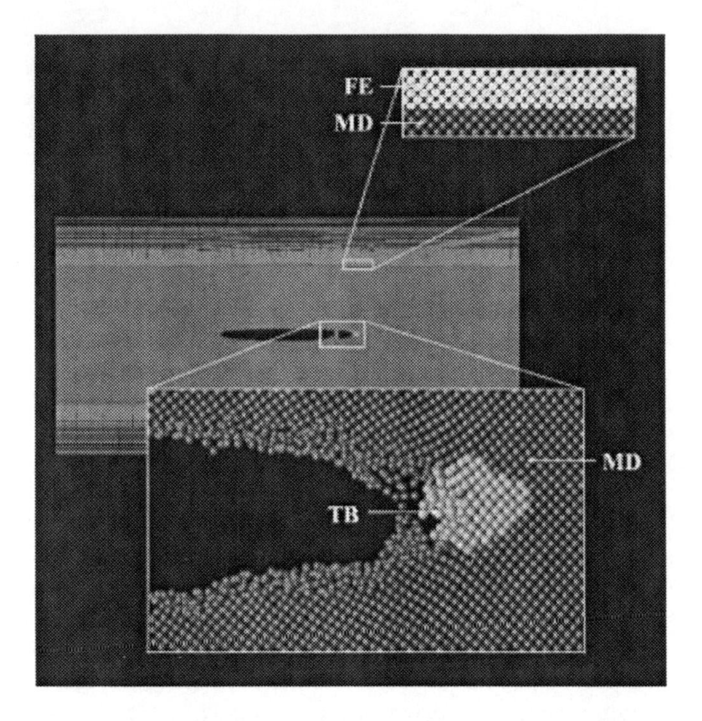

Figure 1. Spatial decomposition into three separate domains. The material is divided into three regions of interest. The yellow region (labeled 'TB') is modeled with the 'Tight Binding' algorithm derived from quantum mechanics. The blue region is modeled using classical molecular dynamics, and the outer, orange region is modeled using a finite element method based on continuum mechanics. (Reprinted with permission from Broughton et al. (1999). Copyright 2006, American Physical Society.)

1999: 2391). What they mean by this is that there is significant feedback between the three regions. All of these effects, each one of which is best understood at its own unique scale of description, are strongly coupled together. Since all of these effects interact simultaneously, it means that all three of the different modeling regions have to be coupled together and modeled simultaneously. The fact that three different theories at three different levels of description need to be employed makes the models 'multiscale.' The fact that these different regions interact simultaneously, that they are strongly coupled together, means that the models have to be 'parallel multiscale.'

An instructive way to think about the meaning of the phrase 'parallel multiscale' is to compare two different ways of going about integrating different scales of description into one simulation. The first more traditional method is what Abraham's group, in keeping with their computational background, label 'serial multiscale.' The idea of serial multiscale is to choose a region, simulate it at the lower level of description, summarize the results into a set of parameters digestible by the higher level description, and then pass those results up to a simulation of the higher level.

But serial multiscale methods will not be effective when the different scales are strongly coupled together:

> There is a large class of problems for which the physics is inherently multiscale; that is, the different scales interact strongly to produce the observed behavior. It is necessary to know what is happening simultaneously in each region since one is strongly coupled to another (Broughton et al. 1999: 2391).

What seems to be required for simulating an inherently multiscale problem is an approach that simulates each region simultaneously, at its appropriate level of description, and then allows each modeling domain to continuously pass relevant information back and forth between regions – in effect, a model that seamlessly combines all three theoretical approaches. Sticking to language borrowed from computer science, Abraham's group refers to this method as 'parallel multiscale' modeling. They also refer to it as the 'concurrent coupling of length scales.' What allows the integration of the three theories to be seamless is that they overlap at the boundary between the pairs of regions. These boundary regions are where the different regions 'shake hands' with each other. The regions are called the 'handshaking regions' and they are governed by 'handshaking algorithms.' We shall see how this works in more detail in the next section.

The use of these handshaking algorithms is one of the things that make these parallel multiscale models interesting. Parallel multiscale modeling, in particular, appears to be a new way to think about the relationship between different levels of description in physics and chemistry. Typically, after all, we tend to think about relationships between levels of description in mereological terms: A higher level of description relates to a lower level of description more or less in the way that the entities discussed in the higher level are made up out of the entities found in the lower level. That kind of relationship, one grounded in mereology, accords well with the relationship that different levels of models bear to each other in what the Abraham group label serial multiscale modeling. But parallel multiscale models appear to be a different way of structuring the relationship between different levels of description in physics and chemistry.

I would like to offer a little bit more detail about how these models are put together, and, in particular, to say a bit more about how the handshaking algorithms work—in effect, to illustrate how one seamless model can integrate more than one level of description. To do this, though, I have to first of all say a bit more about how each separate modeling level works. I turn to that in the next section.

THREE THEORETICAL APPROACHES

Continuum Mechanics (Linear Elastic Theory)

The basic theoretical background for the model of the largest scale regions is linear elastic theory, which relates, in linear fashion, stress – a measure of the quantity of force on a point in the solid – with strain – a measure of the degree to which the solid is deformed from equilibrium at a point. Linear elastic theory, combined with a set of experimentally determined parameters for the specific material under study, enables you to calculate the potential energy stored in a solid as a function of its local deformations. Since linear elastic theory is continuous, in order for it to be used in a computational model, it has to discretized. This is done using a 'finite element' method. This technique involves a 'mesh' made up of points that effectively tile the entire modeling region with tetrahedra. The size of each tetrahedron can vary across the material being simulated according to how much detail is needed in that area (see the bottom of Figure 2.a). Each mesh point is associated with a certain amount of displacement—the strain field. At each time step, the total energy of the system is calculated by 'integrating' over each tetrahedron. The gradient of this energy function is used to calculate the acceleration of each grid point, which is, in turn, used to calculate its position for the next time step. And so on.

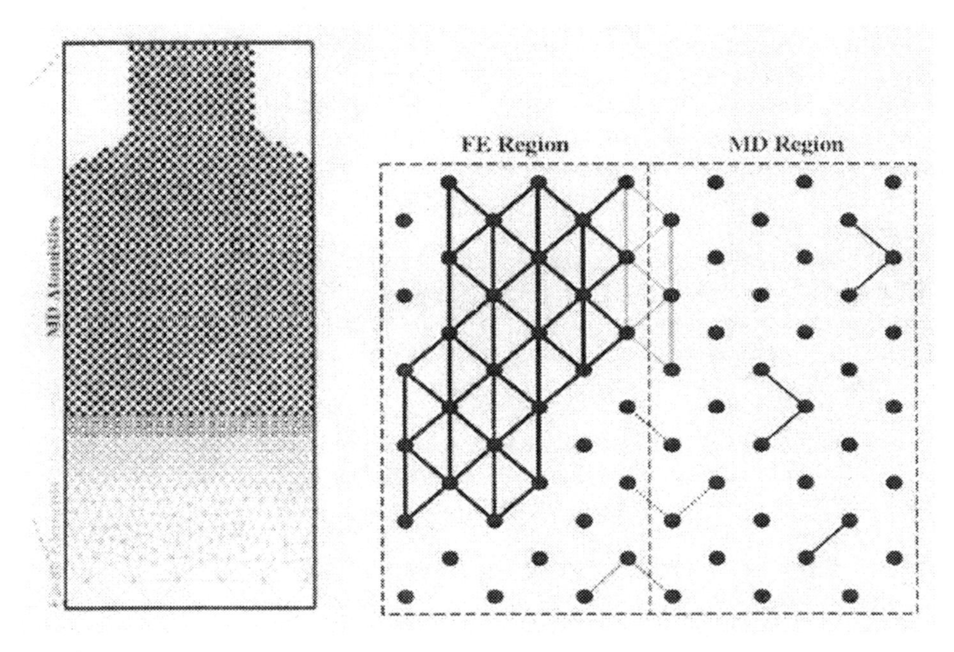

Figure 2: Handshaking. a) (left) To handshake between FE and MD, the size of the FE grid is gradually brought into line with the spacing of the molecules in the MD region. b) (right) When the molecules and mesh points line up, the combined Hamiltonian can be calculated. The dotted lines represent energy values calculated using the MD algorithm. The gray lines represent values calculated using the FE algorithm. The two values are then averaged to come up with the energy of interaction between the two regions. (Reprinted with permission from Abraham et al. (1998). Copyright 2006, American Institute of Physics.)

Molecular Dynamics

In the medium-scale regions, the basic theoretical background is a classical theory of interatomic forces. The model begins with a lattice of atoms. The forces between the atoms come from a classical potential energy function for silicon proposed by Stillinger and Weber (Stillinger and Weber 1985). The Stillinger-Weber potential is much like the Leonard-Jones potential in that its primary component comes from the energetic interaction of nearest neighbor pairs. But the Stillinger-Weber potential also adds a component to the energy function from every triplet of atoms, proportional to the degree to which the angle formed by each triplet deviates from its equilibrium value. Just as in the finite element case, forces are derived from the gradient of the energy function, which are, in turn, used to update the position of each atom at each time step.

Quantum Mechanics

The very smallest regions of the solid are modeled as a set of atoms whose energetic interaction is governed, not by classical forces, but by a quantum Hamiltonian. The quantum mechanical model they use is based on a semi-empirical method from computation quantum chemistry known as the 'Tight-Binding' method. It begins with the Born-Oppenheimer approximation. This approximation separates electron motion and nuclear motion and treats the nuclei as basically fixed particles as far the electronic part of the problem is concerned. The next approximation is to treat each electron as basically separate from the others, and confined to its own orbital. The semi-empirical part of the method is to use empirical values for the matrix elements in the Hamiltonian of these orbitals. For example, the model system that Abraham's group has focused on is solid-state silicon. So, the values used for the matrix elements come from a standard reference table for silicon – derived from experiment. Once again, as soon as a Hamiltonian can be written down for the whole system, the motions of the nuclei can be calculated from step to step.

HANDSHAKING BETWEEN THEORIES

Clearly, these three different modeling methods embody mutually inconsistent frameworks. They each offer fundamentally different descriptions of matter, and they each offer fundamentally different mathematical functions describing the energetic interactions among the entities they describe. "The overarching theme is that a single Hamiltonian is defined for the entire system" (Broughton et al. 1999: 2393).
The key to building a single coherent model out of these three regions is to find the right handshaking algorithm to pass the information about what is going on in one region that will affect a neighboring region into that neighbor. One of the difficulties that beset earlier attempts to exchange information between different regions in multiscale models was that they failed, badly, to conserve energy. The key to Abraham's success in avoiding this problem is that his group constructs their handshaking algorithms in such as way as to define a single expression for energy for the whole system. The expression is a function of the positions of the various 'entities' in their respective domains, whether they be mesh elements, classical atoms, or the atomic nuclei in the quantum mechanical region.

The best way to think of Abraham's handshaking algorithms then, is as an expression that defines the energetic interactions between, for example, the matter in the continuum dynamical region with the matter in the molecular dynamical regions. But this is a strange idea indeed – to define the energetic interactions between regions – since the salient property possessed by the matter in one region is a (strain) field value, whereas the other is the position of a constituent particle, and, in the third, it is an electron cloud configuration. To understand how this is possible, we have to simply look at the details in each case.

Handshaking Between CM and MD

To understand the CM/MD handshaking algorithm, first envision a plane separating the two regions. Next, recall that in the finite element method of simulating linear elastic theory, the material to be simulated is covered in a mesh that divides it up into tetrahedral regions. One of the original strengths of the finite element method is that the finite element mesh can be varied in size to suit the simulation's needs, allowing the simulationists to vary how fine or coarse the computational grid is at different locations. When the finite element method is being used in a multiscale model, this feature of the FE mesh becomes especially useful. The first step in defining the handshake region is to ensure that as you approach the plane separating the two domains from the FE side, the mesh elements of the FE domain are made to coincide with the atoms of the ME domain. (Farther away from the plane, the mesh will typically get much coarser.)The next step is to calculate the energy of the 'handshake region.' This is the region between the last mesh point on one side and the first atom on the other. The technique that Abraham's group use is essentially to calculate this energy twice – once from the perspective of FE, and once from the perspective of MD – and then average the two. Doing the first of these involves pretending that the first row of atoms are actually mesh elements; doing the second involves the opposite – pretending that the last row of mesh element are atoms (see Figure 2).

Suppose, for example, that there is an atom on the MD side of the border. It looks over the border and sees a mesh point. For the purpose of the handshaking algorithm, we treat that mesh point as an atom, calculate the energetic interaction according to the Stillinger-Weber potential, and we divide it by two (remember, we are going to be averaging together the two energetics). We do this for every atom/mesh-point pair that spans the border. Since the Stillinger-Weber potential also involves triples, we do the same thing for every triple that spans the border (again dividing by two). This is one half of the 'handshaking Hamiltonian.' The other half comes from the continuum dynamics' energetics. Whenever a mesh point on the CM side of the border looks over and sees an atom, it pretends that atom is a mesh point. Thus, from that imaginary point of view, there are complete tetrahedra that span the border (some of whose vertices are mesh points that are 'really' atoms.) Treating the position of that atom as a mesh-point position, the algorithm can calculate the strain in that tetrahedron, and integrate over the energy stored in the tetrahedron. Again, because we are averaging together two Hamiltonians, we divide that energy by two.

We now have a seamless expression for the energy stored in the entire region made up of both the continuous solid and the classical atoms. The gradient of this energy function dictates how both the atoms and the mesh points will move from step to step. In this way, the happenings in the CM region are automatically communicated to the molecular dynamics region, and vice versa.

Handshaking Between MD and QM

The general approach for the handshaking algorithm between the quantum region and the molecular dynamics region is similar: The idea is to create a single Hamiltonian that seamlessly spans the union of the two regions. But, in this case, there is an added complication. The difficulty is that the tight-binding algorithm does not calculate the energy locally. That is, it does not apportion a value for the energy for each interatomic bond; it calculates energy on a global basis. Thus, there is no straightforward way for the handshaking algorithm between the quantum and MD region to calculate an isolated quantum mechanical value for the energetic interaction between an outermost quantum atom and a neighboring innermost MD atom. But it needs to do this in order to average it with the MD value for that energy.

The solution that Abraham and his group have developed to this problem is to employ a trick that allows the algorithm to localize that QM value for the energy. The trick is to employ the convention that at the edge of the QM region, each 'dangling bond' is 'tied off' with an artificial univalent atom. To do this, each atom location that lies at the edge of the QM region is assigned an atom with a hybrid set of electronic properties. In the case of silicon, what is needed is something like a silicon atom with one valence electron. These atoms, called 'silogens,' have some of the properties of silicon and some of the properties of hydrogen. They produce a bonding energy with other silicon atoms that is equal to the usual Si-Si bond energy, but they are univalent like a hydrogen atom. This is made possible by the fact that the method is semi-empirical, and so fictitious values for matrix elements can simply be assigned at will. This makes it such that the silogen atoms do not interact energetically with their silogen neighbors, which means that the algorithm can localize their quantum mechanical energetic contributions. Finally, once the problem of localization is solved, the algorithm can assign an energy between atoms that span the threshold between regions that is the average of the Stillinger-Weber potential and the energy from the Hamiltonian in the tight-binding approximation. Again, this creates a seamless expression for energy.

THREE QUESTIONS

In the sequel, I shall suggest that there are features of these multiscale models – with their integration of different levels of dscription, their 'handshaking algorithms,' and their silogens – that appear on their face to be at odds with some basic philosophical intuitions about the relationships between different theories and between theories and their models. But before I begin to draw any philosophical conclusions, I think it is important to note that this area of research – nanomechanics in general and these multiscale methods in particular – is in its relative infancy. And while Abraham and his group have had some success with their models, researchers in these areas are still facing important challenges. It is probably too early to say whether or not this particular method of simulation will turn out, in the great scheme of things, to be the right way to go about predicting and representing the behavior of 'intermediate-sized' samples of solid-state materials. Hence, it is probably also too early to be drawing conclusions, methodological or otherwise, from these sorts of examples.

On the other hand, it might *not* be too early to start thinking about what kinds of basic philosophical intuition about science are likely to come under pressure – or to be informed in novel ways – if and when these scientific domains mature. So we might, at this stage, try to pinpoint some basic philosophical questions; questions whose answers are likely to be influenced by this kind of work. In other words, what I want to do here is simply to offer some ideas about what kinds of question philosophers are likely to be able to shed light on, prospectively, if they keep an eye on what is going on in nanoscale modeling and simulation – especially with regard to multiscale metods – and to provide a sneak preview of what we might discover as the field progresses. Here are three such questions:

(1) *What relationships are possible between levels of description?*

One issue receiving perennial attention from philosophers of science has been that of the relationship between different levels of description. Traditionally, the focus of this inquiry has been debate about whether or not, and to what extent or in what respect, laws or theories at higher levels of description are reducible to those at a lower level.

Underlying all of this debate, I believe, has been a common intuition: The basis for understanding interlevel interaction – to the extent that it is possible – is just applied mereology. In other words, to the extent that the literature in philosophy of science about levels of description has focused on whether and how one level is reducible to another, it has implicitly assumed that the only interesting possible relationships are logical ones –that is, intertheoretic relationships that flow logically from the mereological relationships between the entities posited in the two levels.[2]

But if methods that are anything like those described above become accepted as successful in nanoscale modeling, that intuition is likely to come under pressure. The reason is that so-called ‘parallel’ multiscale modeling methods are forced to develop relationships between the different levels that are perhaps suggested, but certainly not logically detemined, by their mereology. Rather, developing the appropriate relationships, in Abraham’s words, “requires physical insight.”

What this suggests is that there can be a substantial physics of interlevel interaction; a physics that is guided, but by no means determined, by either the theories at each level or the mereology of their respective entities. Indeed, whether or not the relationships employed by Abrahan and his group will turn out to be the correct ones is an empirical/physical question and not a logical/mereological one.

(2) *How important is the consistency of a set of laws?*

This is an issue that has begun to receive attention only recently, particularly in the work of Mathias Frisch (2004). Using classical electrodynamics (CED) as an example, Frisch has challenged a common philosophical intuition about scientific theories: that the internal consistency of their laws is a necessary condition that all successful theories have to satisfy. I want to make a similar point here. In this case, the example of multiscale modeling seems to put pressure on a closely related, if somewhat weaker, intuition: that an inconsistent set of laws can have no models.

In a formal setting, this claim is obviously true; indeed, it is true by definition. But rarely in scientific practice do we actually deal with models that have a clear formal relationship to the laws that inspire them. Most likely, the intuition that inconsistent laws cannot produce a coherent model in everyday scientific practice rests as much on pragmatic considerations as it does on the analogy to formal systems: How, in practice, could mutually conflicting sets of laws guide the construction of a coherent and successful model?

We can start by looking at what we learn from Frisch. In CED, the strategy is usually to keep the inconsistent subsets of the theory properly segregated for a given model.

> The Maxwell-Lorentz equations can be used to treat two types of problem. We can appeal to the Maxwell equations to determine the fields associated with a given charge and current distribution; or we can use the Lorentz force law to calculate the motion of a charged particle in a given external electromagnetic field (Frisch 2004: 529).

In other words, in most models of CED, each respective model draws from only one of the two mutually inconsistent 'sides' of the theory. This technique works for most applications, but there are exceptions in which the method fails. Models of synchrotron radiation, for example, necessarily involve both mutually inconsistent parts of the theory.

There are problems, in other words, that require us to calculate the field from the charges, as well as to calculate the motion of the charges from the fields. But the solution method, even in the synchrotron case as Frisch describes it, is still a form of segregation. The segregation is temporal. You break the problem up into time steps: In one time step, the Lorentz equations are used; in the next, the Maxwell equations; and so on.

A form of segregation is employed in multiscale modeling as well, but it is forced to break down at the boundaries. Each of the three theoretical approaches is confined to its own spatial region of the system. But the fact that there are significant simultaneous and back-and-forth interactions between the physics in each of these regions means that the strategy of segregation cannot be entirely effective. *Parallel* multiscale methods require the modeler to apply, in the handshaking region, two different sets of laws. The laws in Abraham's model, moreover, are each pair-wisely inconsistent. They offer conflicting descriptions of matter and conflicting accounts of the energetic interactions between the constituents of that matter. But the construction of the model in the handshaking regions is guided by both members of the pair. When you include the handshaking regions, parallel multiscale models are – all at once – *models of an inconsistent set of laws*

The methods developed by these researchers for overcoming these inconsistencies (the handshaking algorithms) may or may not turn out to be too crude to provide a reliable modeling approach. But by paying close attention to developments in the field of nanoscale modeling, a field in which the models are almost certainly going to be required to involve hybrids of classical, quantum, and continuum mechanics, philosophers are likely to learn a great deal about how inconsistencies are managed. In the process, we shall be forced to develop richer accounts of the relationships between theories and their models – richer accounts, in any case, than the one suggested by the analogy to formal systems.

(3) *How do models differ from ideal descriptions?*
(What role can falsehoods play in model building?)

It has been widely recognized that many successful scientific models do not represent exactly. A simple example: The model of a simple harmonic oscillator can quite successfully predict the behavior of many real physical systems, but it provides at best only an approximately accurate representation of those systems. Nevertheless, many philosophers hold to the intuition that successful models differ from ideal descriptions primarily in that they include idealizations and approximations. Ronald Laymon has made this intuition more precise with the idea of "piecewise improvability" (Laymon 1985). The idea is that while many empirically successful models deviate from ideal description, a small improvement in the model (i.e., a move that brings it closer to an ideal description) should always result in a small improvement in its empirical accuracy.

But what about the inclusion of 'silogens' in multiscale models of silicon? Here, piecewise improvability seems to fail. If we make the model 'more realistic' by putting in more accurate values for the matrix elements at the periphery of the QM region, then the resulting calculation of the energetic interactions in the handshake region will become less accurate, not more accurate, and the overall simulation will fail to represent accurately at all. The lesson of this and other examples is that models can sometimes successfully make use not only of approximations and idealizations but also outright 'falsifications.' False assumptions, it appears, can be systematically successful.[3] Nanoscale models, particularly simulation models, are likely to put pressure on the philosophical intuition that success and reliability always come from truth.

ACKNOWLEDGMENTS

I thank the many attendees at PSA 2004 and at a presentation at the University of Bielefeld for helpful comments.

University of South Florida, Tampa, FL, USA

*Reprinted with permission from E. Winsberg "Handshaking Your Way to the Top: Simulation at the Nanoscale" (enlarged version), *Philosophy of Science*, PSA 2004; to appear. Copyright 2006 by University of Chicago Press.

NOTES

[1] Good review literature on parallel multiscale simulation methods for nanomechanics can be found in Abraham et al. 1998; Broughton et al. 1999; and Rudd and Broughton 2000.

[2] An important exception is the recent work of Robert Batterman (2002).

[3] For other examples of 'falsifications,' as well as for a discussion of their implications for scientific realism, fundamentalism, and the status of 'reliability' as a viable semantic notion, see Winsberg (forthcoming).

REFERENCES

Abraham, F.F., J.Q. Broughton, N. Bernstein, and E. Kaxiras (1998). "Spanning the length scales in dynamic simulation", *Computers in Physics*, **12** (6): 538–546

Batterman, R. (2002). The Devil in the Details: Asymptotic Reasoning in Explanation, Reduction, and Emergence, New York: Oxford University Press.

Broughton, J., F.F. Abraham, N. Bernstein, and E. Kaxiras (1999). "Concurrent coupling of length scales: Methodology and application", *Physical Review B*, **60** (4): 2391–2403.

Chelikowski, J. and M. Ratner (2001). "Nanoscience, nanotechnology, and modeling", *Computing in Science and Engineering*, **3** (4): 40–41.

Frisch, M. (2004). "Inconsistency in classical electrodynamics", *Philosophy of Science*, **71**(4): 525–549.

Laymon, R. (1985). "Idealization and the testing of theories by experimentation", in P. Achinstein and O. Hannaway (eds.), *Observation, Experiment and Hypothesis in Modern Physical Science*, Cambridge, MA: MIT Press, pp. 147–173.

Nakano, A., M.E. Bachlechner, R.K. Kalia, E. Lidorikis, P. Vashishta, G.Z. Voyiadjis, T.J. Campbell, S. Ogata, and F. Shimojo (2001). "Multiscale simulation of nanosystems", *Computing in Science and Engineering*, **3** (4): 56–66.

Rudd, R.E. and J.Q. Broughton (2000). "Concurrent coupling of length scales in solid state systems", *Physica Status Solidi B*, **217**: 251–291.

Stillinger, F.H. and T.A. Weber (1985). "Computer simulation of local order in condensed phases of silicon", *Phyiscal Review B*, **31**: 5262–5271.

Winsberg, E. (forthcoming). "Models of success vs. the success of models: Reliability without truth", *Synthese*.

SOCIAL PRACTICE

CHAPTER 10

MARTINA MERZ

LOCATING THE DRY LAB ON THE LAB MAP

Virtual laboratory, digital laboratory, dry lab – notions such as these frequently become associated with computer simulation be it in popular accounts or in the discourse of practitioners. The notions seem to suggest that computers and simulation applications constitute research environments in their own right, allowing one to perform computer experiments and endowing one with the potential to replace traditional laboratory settings. The images carry connotations about the kind of work involved and the work settings in which scientists and their objects of investigation relate to each other. The present chapter takes up these ideas with the understanding that they prompt important questions for philosophers, historians, and social scientists of simulation practice. The chapter is an attempt to come to grips with some of these questions by exploring the association between *simulation* and the *laboratory*.

A first hypothesis that one may associate with the image of a dry lab posits that simulation (and modeling more generally) is *like experimental work* in allowing scientists to explore processes, manipulate structures, and tinker with parameters and conditions. Science studies authors such as Dowling (1999), Keller (2003), Morgan (2003), and Sismondo (1999) have argued that this practice is indeed typical of simulation and modeling. While to associate simulation with experiment in such a way brings out the methodological features of simulation practice, to view simulation as constituting a laboratory of its own raises additional questions and hypotheses that have received little attention in the literature as yet. What then comes into view are issues to do with the simulation laboratory's relation to its boundaries and to objects and practices 'outside' the lab, which also brings up the issue of the laboratory's autonomy. The perspective raises the question how, to what extent, and to what effect simulation activities either become disconnected from other scientific practices or remain associated with them.[1] This article's *central thesis* asserts that to adequately conceptualize simulation one has to address not only the disembedded nature of simulation but also the *dynamic interplay of the disembedding and reembedding* moves that determine how and to which epistemic means simulation is employed.

In what sense then can – or should – computer simulation be interpreted as constituting a laboratory? To explore this issue, I shall draw on laboratory concepts that have developed in the social studies of science (next section). The chapter argues that

J. Lenhard, G. Küppers, and T. Shinn (eds.), Simulation: Pragmatic Construction of Reality, 155–172.

different laboratory notions allow one to reveal different features of simulation work, each concept defining a distinct frame for the analysis of computer simulation practice. This approach results in a multilevel account of simulation practice that will be unraveled by discussing a specific case: computer simulation in particle physics. I conducted the ethnographic study on which this analysis is based over a period of approximately ten years at CERN, the European Laboratory for Particle Physics in Geneva (Switzerland). Particle physics provides a particularly interesting case due to the great significance, the extensive application, and the multiple roles that simulation assumes in this research area (see also Merz 1999).

The present account of simulation practice operates on several levels. To provide a readable narrative, the levels cannot be disentangled entirely. They will be presented alongside one another with an effort to render them explicit throughout the text. After introducing the reader to the case study of particle physics, two laboratory concepts will guide the discussion on simulation practice, each presented in a special section. The levels of analysis concern, first of all, these two perspectives: the *configuration of objects* and the *local intertwining* of simulation with other practice. Furthermore, simulation and other practices are interlaced in time, which brings to the fore a *temporal dimension* that is of particular importance in the case of particle physics. Finally, besides providing a close reading of the epistemic practice of simulation, the chapter pursues a *reflexive aim*. It raises the question in what sense accounts of scientific practice rely on the underlying conceptual frameworks. In a way, then, this chapter performs an experiment, its experimental target being the extended case study of simulation at CERN.

CONCEPTS OF THE LABORATORY

The science-as-practice approach of the new sociology of science has developed laboratory concepts that are productive for the present exploration of simulation. I distinguish two complementary characterizations of the laboratory, which have both been elaborated by the same set of authors and represent different focal points and targets of argumentation.

Laboratory as a Site of Locally Embedded Practice

The first perspective views laboratory research as being inextricably tied to the locales in which knowledge is produced (see for an overview, e.g., Lynch 1997: chap. 3). The laboratory is seen as a repository of competencies, tools, and resources that the scientists draw upon. Scientists exploit the contingencies of local contexts with respect to the equipment and research facilities at hand, the interactional circumstances, the conventions embodied in laboratories, the combined expertise gathered in a research team, and the organizational setting in which it is embedded. Scientists employ a whole repertoire of improvisations and tentative solutions, different forms of tinkering and embodied skills, as well as different techniques of persuasion and negotiation. As a consequence, the research problems are locally constituted, as are the research objects, the tools, and the way in which scientists handle and assemble all these elements. Out of this seemingly messy set of things and actions, scientists "produce order" (Latour and Woolgar 1979), which points to the constructive nature of scientific work. In this perspective, knowledge production is closely associated with the laboratory as a site of locally embedded practice.

Laboratory as a Site of Object Reconfiguration

A second perspective identifies the laboratory with a lieu in which objects (and subject-object relations) are reconfigured and acted upon in specific ways. As the paramount site of knowledge production in modern science, the lab symbolizes and substantiates the power and success of science. Bruno Latour and Karin Knorr-Cetina, among others, maintain convincingly that this power relies on specific forms of object work that are performed in – and are constitutive of – the laboratory.

Bruno Latour (1983)[2] argues that scientists gain strength in the laboratory by inverting the hierarchy of forces according to their research interests. They do this by reversing the scale of phenomena in the laboratory, making some objects bigger and others smaller. For example, organisms are isolated and cultivated in a suitable milieu, which allows them to grow exponentially and become visible to the scientist's eye. As a consequence, scientists are enabled to do things in the lab that are not feasible outside where the existing scales are unmanageable. The variation of scales allows scientists also to multiply experiments at reduced cost, which turns the laboratory into a learning environment, "a technological device to gain strength by multiplying mistakes" (Latour 1983).

Karin Knorr-Cetina (1992) similarly argues that the laboratory is "an enhanced environment" (Knorr-Cetina 1992: 116). She identifies the mechanism that brings this about as the reconfiguration of subject-object relations to the scientists' advantage, which can be viewed as a generalized notion of Latour's scale reversal. In the laboratory, scientists reshape the phenomena of investigation in order to control their temporal and spatial accessibility and to render them fit for experimentation. Lab objects can be duplicated, standardized, and are amenable to a full sequence of experiments. In addition, social relations are also reconfigured and aligned with the specific requirements of the objects in the lab. For example, collaborations are forged to confront the object world optimally, with the form and size of collaborations differing widely across fields. To summarize this perspective: According to the cited literature by Latour and Knorr-Cetina, knowledge production is closely associated with a specific mode of relations between the scientists and their objects that characterize the laboratory.

Both perspectives on the laboratory – the local practice variant and the object reconfiguration variant – can be applied productively to a discussion of simulation practice. The remainder of this chapter will first introduce the case study of particle physics. It will then draw out the specific features of simulation by focusing on each perspective consecutively. Each perspective gives rise to an account with its own legitimacy and logic, emphasizing important elements of simulation practice while downplaying others.

PARTICLE PHYSICS AND SIMULATION

Over the last decades, computer simulation has become a central strategy for the conduct of experiments in high-energy physics. Physicists explain the indispensability of extended simulation work by referring to the specificities of the physics processes (i.e., the statistical nature of the events and the detection process) and the experiments' complexity. The dynamics of ever more powerful beam energies that require large-scale technologies and result in increasingly strong data fluxes also characterize the accelerator experiments at CERN, the European Laboratory for Particle Physics. Consider the case of the Large Hadron Collider (LHC), which was approved by CERN's Council in 1995 and is currently under construction. Its two proton beams will be ready for collision in 2007, enabling physicists to investigate particle properties that have been inaccessible to experiment before. Physicists expect that the LHC will allow them to tackle some of the central unsolved questions of the Standard Model (e.g., related to the Higgs mechanism, supersymmetry, antimatter). Centerpiece of the LHC is its accelerator ring of 27 km circumference that will use superconducting magnets to keep the proton beams on track. The LHC aims at providing the highest energy of accelerators and the most intense beams worldwide.

Collaborations, Detectors, and Simulation

Physicists will perform measurements at different sites along the accelerator ring. At present, five experimental 'collaborations' (ATLAS, CMS, etc.) are installing their measuring devices: the particle detectors. A collaboration is defined by and 'built around' a particular detector. It is responsible for planning, designing, and building the device and, once the experiment is up and running, for taking and analyzing the data it provides. Physicists at CERN refer to the collaborations also simply as 'the experiments.' The particle detectors have the size of multistory buildings. They are made up of several layers (the different detector systems), which, like a set of Russian dolls, are arranged around the collision point. From the innermost to the outermost layer, the particles traverse the tracking chamber, the electromagnetic calorimeter, the hadron calorimeter, and, finally, the muon chamber. Each layer measures different properties of the passing particles, such as particle type, electrical charge, or momentum. The different particles can be identified as each leaves a specific 'signature' in the various detector layers. Electrons and protons, for example, leave a trace in both the tracking chamber and the electromagnetic calorimeter whereas neutrons do not. They, for their part, are detected by the energy they deposit in the hadron calorimeter.

The full lifecycle of the current collider experiments, which encompasses several phases, will extend over fifteen to twenty years.[3] The corresponding collaborations are of considerable size: ATLAS, one of the LHC collaborations, involves more than 2,000 members from thirty-four countries. The engaged competencies comprise traditional skills of the experimenter (e.g., design of instruments, manipulation of apparatus, data analysis, and interpretation) but also include expertise in areas as diverse as management and licensing, electronics and information technology, theoretical particle physics and solid state physics, material sciences and mathematics, and so forth. The collaboration members are sorted into different organizational units, which loosely correspond to what the physicists call different 'communities' (see also Knorr-Cetina 1995). These units or communities are dedicated on the one hand to the aforementioned detector systems (such as the electromagnetic calorimeter) and on the other hand to coordination tasks of concern to the experiment's overall operation (such as physics coordination, computing coordination, or electronics coordination). With respect to this structure, simulation is a *transversal practice*: Simulation work is performed by members of different communities and in different organizational units; it becomes coordinated across units; and it assists in coordinating and synchronizing other type of work throughout different communities.[4] Three types of community are of particular interest to the present account: the 'phenomenologists,' the 'physics community,' and the 'detector communities.'

The *phenomenologists* are not part of an experiment's collaboration. They rather form a subgroup of theoretical particle physicists. Theorists and experimentalists constitute separate scientific communities that evolve in distinct organizational structures. Each has its own research agendas, career patterns, and publication practices. Compared with other theoretical particle physicists (e.g., mathematical physicists), the phenomenologists interact and cooperate more closely with experimental physicists. They explore model extensions in correspondence with the feasibility of experimental studies, they develop simulation programs that will be used by experimentalists later on, they perform simulation studies whose results are of interest to experiment, and so forth.

Among the members of a collaboration, simulation work features importantly both in the 'physics community' and in the different 'detector communities.' The *physics community* is responsible for all aspects of physics analysis, which varies throughout the different phases of an experiment: It specifies the physics priorities of an experiment, guides the optimization procedure for the apparatus to assure that the physics goals can be addressed by the experiment, develops strategies to analyze the data, and, once the experiment is running, performs the data analysis. Simulation is an indispensable instrument to address each one of these tasks. In the different phases, the physics community interacts intensely with phenomenologists on the one hand and with the different detector communities on the other hand.

The *detector communities*, each dedicated to one of the detector systems, are in charge of developing and constructing the different detector systems. In close interaction with the physics community, they have to assure that the detector systems will meet the requirements that follow from the physics priorities. The systems may be broken down into further subunits that correspond to subdetectors. For example, the inner detector is made up of the pixel detector, the semiconductor tracker, and a transition radiation tracker. The detector communities are responsible for the apparatus during its entire lifetime: starting with the design and proceeding to the construction, the maintenance, and finally the working of the respective detector systems. A central task of these endeavors is to reach an understanding of the detector and its operating mode that is as complete and as accurate as possible. Simulation strongly supports detector design work; it is employed for understanding the performance of the detector systems and the complete detector. Simulation expertise and other types of expertise go hand in hand and are coordinated closely to accomplish the tasks of the detector communities.

Finally, the *computing community* should be at least mentioned. It provides the information technology infrastructure that sustains the other communities' respective simulation work. It deals with computer platforms, builds releases, and ensures that the computing efforts throughout the collaboration can be coordinated and synchronized.

Simulating Events and Detector Performance

Simulation is employed to mimic the processes of production and detection of elementary particles as they are thought to occur in a 'real' experiment, with separate computer programs assuming the roles of collider and of detector (see Figure 1). Particle physicists label experiments, data, and so forth as 'real' in contrast to their 'simulated' counterparts.

An *event generator* – like a collider – brings particle pairs into collision, a process in the course of which a plethora of new particles (up to several thousand in the case of the LHC) is created. The program produces collisions one by one, as in the 'real' experiment. Each collision is called an 'event.' The program generates many events in a single computational session; all are produced with the same initial parameters (e.g., the beam energy). What makes each event unique is the final output: the listing of the produced particles, their properties, and their 'history' (i.e., information about the 'mother particles' from which they emerged). The events are different because the laws of quantum physics are probabilistic. The main event generators have been constructed and are maintained by phenomenologists.

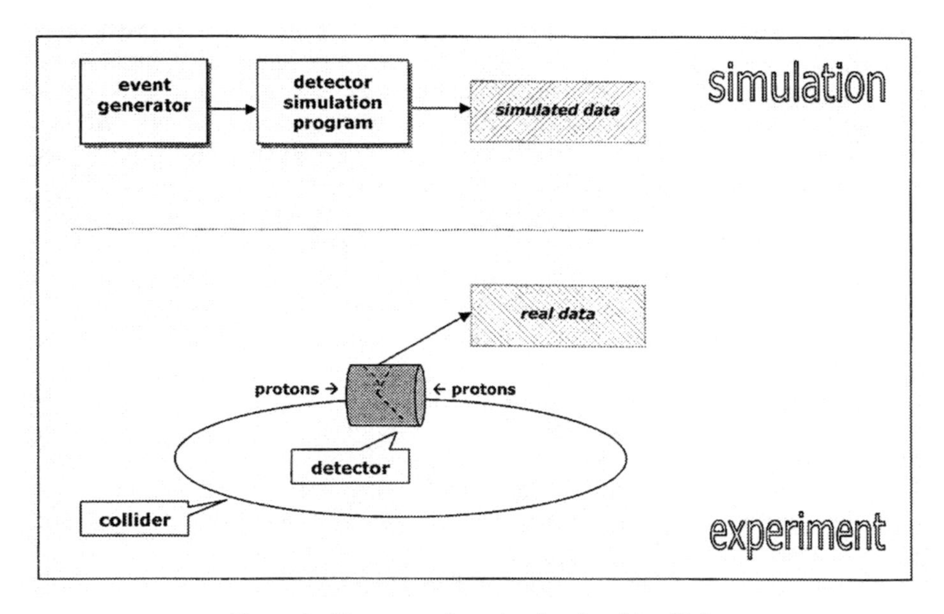

Figure 1. How to produce simulated and 'real' data

A *detector simulation program* accounts for the working of a particle detector with its different detector systems. The experiments at CERN and most other particle physics experiments worldwide use the GEANT program package for this purpose. Designed originally for application in high-energy physics (HEP) experiments, it is now also applied in research fields outside of HEP, for example, in medical and biological sciences, radioprotection, and astronautics. GEANT simulates the passage of elementary particles through matter. The program describes the 'tracking' of particles (i.e., their transport) through an experimental setup in order to simulate the detector response. It also provides for the graphical representation of the experimental setup and the resulting particle trajectories. Since GEANT is a very general framework, it needs to be adapted to an experiment's specific detector. For this purpose, its design and setup are described by members of the detector communities as a structure of geometrical volumes, each volume being characterized by how the incoming particles interact with the detector material. As output, the detector simulation program produces particle trajectories and the response of the detectors.

When coupling the two simulation programs by feeding the output of the event generator into the detector simulation program as its input, the entire event chain – production of elementary particles, collision, detection of the emerging particles – can be modeled in the computer. The output of the simulation runs consists of the 'simulated data.' Simulated data and 'real data' are indistinguishable in their form: The detector simulation program provides output in a format that is identical to that of the experimental data acquisition system. This is a prerequisite for simulation being able to mirror (parallel, substitute, etc.) the real experiment. In the phase of data analysis, for example, the identity of the data formats allows both sets of data to be passed through the same 'reconstruction program' that then enables the comparisons in the process of physics analysis.

Simulated data are produced and processed in dedicated simulation studies throughout all phases of the experiment. The simulation activities are highly varied, be it in their radius of application, their social organization, their epistemic function, the problems they address, or the consequences of their results (see also Merz 1999). Some of this rich texture will be exposed in the next two sections. The aforementioned perspective on the laboratory as identified with the idea of object reconfiguration will feature first.

SIMULATION AS OBJECT RECONFIGURATION

In the wake of the cited work by Latour and Knorr-Cetina, science studies scholars have argued that the reconfiguration of objects in the laboratory can assume different forms. Whereas object reconfiguration in the lab typically modifies the objects' 'material' shape, the concept can also be extended to 'immaterial' forms of reconfiguration. Karin Knorr-Cetina (1999: 27) provides an example from astronomy. She considers the digitized recordings of astronomical phenomena that are transferred to computer terminals where they can be processed independently from the temporal and spatial constraints of a field science. In a related sense, the notion of object reconfiguration can be extended to the case of computer simulation. Phenomena that are numerically and mathematically configured (or modeled) become amenable to experimentation by way of simulation studies. Experimenting with numerically configured phenomena requires an adequately furnished environment: It relies on the availability of appropriate computer models (event generators, detector simulation programs) as well as on a set of competencies, tools, and infrastructure needed to bring simulation studies to a satisfactory conclusion. In a way, the computer appears as a functional equivalent to the workbenches of a traditional laboratory science. In this sense, simulation studies can be viewed as being performed in a digital laboratory. Simulation allows scientists to represent, shape, and experiment on natural, technical, or formal processes and phenomena such as natural systems or research apparatus. Simulation studies enable scientists to mimic complex object environments, and to model and analyze the respective dynamic evolutions. Scientists exploit these options in multiple ways, for example, by exploring new spaces of action and probing their limitations. How they go about this work will be illustrated and analyzed in the following by focusing successively and in chronological order on the different phases of particle physics experiments (see Table 1).

Table 1. Simulation in the life of an experiment

Chronology	**Aim of simulation studies**	**Simulation laboratory**
Before decision to build new machine:	– explore accelerator potential – develop analysis strategies	simulating (possible) accelerators
After approval of new machine:	– design and optimize detector – revise analysis strategies	simulating (possible) detectors
Experiment is running (expected 2007):	– analyze experimental data, explore theoretical models	**Data analysis lab** joint and parallel processing of simulated and real data

Exploring Accelerators

Simulation is relevant in a first, explorative phase of a collider experiment in which the 'discovery potential' of a projected accelerator is tested. In this phase, physicists perform simulation studies to investigate which physics processes will occur as a result of a projected accelerator's particle collisions – the accelerator being characterized by the envisaged particle type, the particle's collision energy, and the beam intensities. In the words of simulation expert Johan,[5] simulation serves at this stage "to estimate the feasibility of an intended physics study." The event generators stand in authoritatively for the – as yet unbuilt – accelerator. The simulation studies constitute a 'dry lab' in which the potential of a future machine to deliver interesting physics processes is tested. Simulation studies have the advantage of allowing physicists to systematically vary and control the machine's parameters in utmost flexibility and to probe for the consequences of the different parameter selections. Event generator expert Paolo asserts: "Event generators are used to do all of the preliminary work that eventually leads to the decision of building a machine." In this phase, the event generator substitutes for the accelerator. The responsibility for these simulation studies lies primarily with the 'phenomenologists' who are engaged in simulation work in a variety of ways. They employ simulation to obtain information on measurable quantities (i.e., the physical properties accessible by experiment) from abstract theories. They perform simulation studies to investigate the theoretical models that underlie the simulation programs. The 'authors' (i.e., producers) of event generators – and the event generators themselves – also act as 'go-betweens' linking theory and experiment, and they mediate between the two epistemic and social realms (see Merz 1999, 2002 and footnote 1).

Devising Analysis Strategies

Once a specific accelerator layout has been judged appropriate for investigating a new type of physics event of particular interest (e.g., a Higgs-particle decaying into two gammas), analysis strategies need to be devised.[6] Devising analysis strategies consists in looking for clever ways to separate the 'signal' from the 'background.' Experimentalists denote as 'background' all those unwanted processes within which they need to uncover the events of interest: the 'signal.' The background needs to be fought and overcome (Knorr-Cetina 1999: 123–127). Exploring ways to 'beat it down' consists in searching for "observables that are optimally discriminating between the signal and the background" (Johan). Such strategies rely on theoretical reasoning, and they are tested for their effectiveness by feeding them into simulation studies. This type of simulation work thus serves to promote theoretical understanding and is performed by members of the physics community: the experts in data analysis.

Firmly embedded in the social life of a collaboration, experts in data analysis use simulation as an indispensable instrument of physics analysis. Their more general aim is to process data in such a way that the result can be compared directly with the predictions that follow from theoretical models and assumptions. The experts in physics analysis participate in a range of physics-simulation working groups, the so-called 'physics performance groups,' each of which is dedicated to a different class of physics events (e.g., Higgs bosons, supersymmetry, B-physics, top physics). The Higgs working group, for example, studies the different decay channels of the – as yet undiscovered – Higgs particle with important decay channels being Higgs to gamma-gamma, Higgs to Z-gamma, and Higgs to four leptons. A considerable number of detailed simulation studies is needed to prepare for the Higgs searches in the (future) actual experiment, and this requires all relevant signatures and various physics scenarios to be taken into account. Physicists base their studies on the 'Standard Model' as well as on as yet experimentally unsupported extensions of it, such as the

'Minimal Supersymmetric Standard Model.' The underlying theoretical framework enters the simulation studies through the choice of a specific event generator whose output is fed into the detector simulation program.

The simulation studies contribute to a better understanding of how evidence for the existence (or inexistence) of the Higgs particle can be extracted from the data. In this phase of the experiment, data analysts work exclusively with simulated data. Their studies are considered as an indispensable preparation for the analysis of future 'real' data. At this stage, event generators are typically employed without coupling them to a detector simulation program. The reason is that both (accelerator and analysis strategies) are probed independently from the detection mechanisms, which will feature next.

Designing Detectors

In a subsequent phase, the design and the layout of the particle detector are of central concern – and simulation studies are devoted to this task. A detector simulation program propagates particles through a (simulated) apparatus to study the apparatus' detection capacity. Physicists scrutinize the detector requirements with the aim of optimizing the design features. What is at stake here is the detector's capacity to be sensitive to the physics events of interest. Experimentalists want to know "how to optimize our detectors in such a way that we could find what we already know today to be interesting physics like, for instance, Higgs going to gamma gamma" (Johan). Different actors in the collaboration negotiate and balance the detector specifications while taking conflicting interests into account. These are not only physics priorities, but also financial and technical constraints. The decisions on design options are taken on the basis of simulation studies. Simulation plays an essential role as a bargaining tool: It enables scientists to flexibly reconsider a variety of design features with the possibility of isolating specific features. It constitutes an enhanced environment that allows physicists to zoom in on contested areas of the detector and test out a variety of potential solutions to problems with the design. In this process, the detector simulation program stands in for the experimental setup of the projected detector systems. The details of the detector design are fixed step by step according to the outcomes of simulation studies.

This simulation work is performed by members of the detector communities who contribute centrally to the conduct of experiment. Each detector community has its own simulation experts. They assume different tasks in different phases of the experiment in line with the research program of the community. In an early phase of designing a detector system, simulation is used, for example, to test out new ideas. Physicist Pat recalls:

> You can use simulation to test an idea in advance and in fact I have done one particular, rather large project. [...] Our barrel silicon tracker is laid out very conventionally. It's just a cylinder with a bunch of flat wafers put around it. But then these people came along from one university and they had a very different idea about how these wafers could be set out, and they thought it was a wonderful idea really. It looked a little funny when you just looked at the drawings on paper. Just looking at it on paper, you'd say that's not going to work. And their claim was, we think it will work and we want to *prove it to you by simulating it* and running some tracks through it. Okay, now this

design was only in their heads and on the paper. But I helped them code it in GEANT and whether they proved that it was better or not doesn't matter [here], but the point is that we made a simulation of an idea that hadn't been built (Pat, interview).

After a considerable stretch of time and highly complex negotiation processes within and between the different communities of a collaboration (which this chapter cannot discuss for reasons of space), the collaboration takes its decision on the detection mechanisms and the detailed detector layout based on the results of simulation studies. At this stage, the aforementioned analysis strategies need to be re-evaluated in relation to the fixed detector specifications. The physics performance groups come to the fore again, and now base their studies on a combined application of event generators and the full detector simulation program, which represents the detector as it will be built.

At this point, the simulated detector layout is stripped of its potential to be adapted to ever new design ideas: it becomes 'frozen' and is declared a central and centering instrument of the collaboration. In large-scale 'productions,' which are organized and supervised by the computing community of the collaboration, a great number of events is produced by extended runs of the official detector simulation program. These data are saved on tape and are rendered accessible to the physics analysis community for detailed analysis.

Running the Experiment

Once the detector has been built and the experiment is running – a stage which (at the time of writing this article) lies in the future of the LHC experiments – simulation no longer needs to stand in for either accelerator or detector as in all previous phases. Again, simulation does not lose its prime importance because new challenges and requirements are to be met. Simulation now provides a parallel set of data for comparison with experimental data. On the basis of this comparison, physicists can decide whether the experimental results agree with the theoretical predictions. For this reason, computer simulation also remains a central instrument in the phase in which the observed data are analyzed and interpreted.

At the end of this chronological itinerary through the phases of an experiment, the underlying perspective on simulation as object reconfiguration can be assessed: The account of the importance and role of simulation throughout different experimental phases may seduce the reader into believing that simulation studies might fully substitute for material exploration until the moment that the 'real' collider and detectors are constructed. The account presents the *fiction of a (materially) disembedded practice*. The simulation laboratory seems to be completely decoupled from the material world of 'traditional' experimentation. Simulation has indeed developed as a distinguished and distinguishable field of expertise in particle physics. In accordance with the increasing specialization that characterizes the social organization of collaborations in particle physics, a tendency to disembed simulation activities entirely from other forms of epistemic practice can be observed. Yet, the conceptual frame that perceives simulation as a particularly successful form of object reconfiguration is at the same time erroneously partial, fragmentary, and fraught with hidden assumptions.

It presents the public imagery of simulation as a valid substitute for (material, concrete) experimental practice, which, in its most radical form, may question the sheer *raison d'être* of concrete experimental practice. Simulation seems to replace *in silico* the rich texture of experimentation. It comes as no surprise that such an account leaves out important aspects of simulation practice. The next section will switch perspective and provide a glimpse of an alternative account.

SIMULATION AS LOCAL PRACTICE

If, in contrast, the laboratory is identified with the multiple instances of local practice, the embedded character of simulation work comes to the fore. In this case, attention is drawn to simulation together with the entire social and material context in which it is embedded and which brings it about. Such a 'wide-angle' approach renders visible the social and technical infrastructure that is needed for performing simulation studies; the epistemic and social entanglement of simulation activities with other endeavors in experimental physics, theoretical physics, or other fields of practice; the relations, rankings, and hierarchies of different data types; and so forth. As a result, what seemed to mark the boundaries of simulation in the first account (see previous section) constitutes the irreducible context of simulation in the second. Depending on the perspective chosen to address simulation work, activities disappear from view in one case to appear as indispensable elements for the pursuit of simulation in another.

Consider the relation between simulation and material practice or 'real data.' In Table 1, which exhibits the different phases of an experiment with the corresponding roles of simulation, real data seem to show up only in the very last stage: once the full experiment is running and producing data. In an important sense, this view is unjustifiably reductionist. This is illustrated in the following by discussing the case of prototyping and test beams that engages the different detector communities.

When physicists develop ideas for detector techniques and devise concepts for its design, its setup, and the principal detection mechanisms on which the final detectors are to be based, simulation is not their only epistemic strategy. The following account fills in the blind spot and focuses on activities that remained concealed when discussing detector design under the perspective of object reconfiguration. Other research sites are set up in parallel and interact with simulation studies in interesting ways. In particular, this work involves physicists who construct and experiment with detector prototypes – in the sense of material artifacts. Prototypes are employed to test detection mechanisms by maximally reducing the complexity and compositeness of the envisaged full detector: Prototypes are miniaturized and simplified versions of the full detector. For example, a prototype may be made up of one module of (what is later on to become) one of the detector systems.

Physicists place detector prototypes in a test setup and expose them to test beams 'to understand how exactly the detector is behaving.' Test-beam studies constitute model and miniature versions of an (entire) experiment. In many cases, physicists propose to use technologies for the new detectors that are not well understood yet, be it that new detector technologies are being explored, or be it that traditional ones are to be employed at considerably higher beam energies than before. For this reason, physicists need to investigate how a (projected) detector will translate physics processes into an electronic signal and measure the detector's resolution. This is done by building prototypes and placing them in test beams. This work is also accompanied by simulation studies. Yet, compared with the cases discussed earlier in which simulation work seemed to precede and substitute for 'real' experimentation, there is an important difference. In the case of prototype studies, simulation and hardware work are *highly interdependent and complementary* forms of knowledge production. The real data produced in prototype experiments are indispensable elements for rendering the simulations of the future detector adequate and thus useful, while simulation studies provide important input for furthering the understanding of the detection mechanisms.

In the phase in which the prototype is probed in test beams, simulation and hardware work are performed in close cooperation and alignment within a detector community. A unit of ten to fifteen (and sometimes more) scientists is responsible for simulating, constructing, and testing the prototype embedded in the wider context of a detector system community. A simulation expert writes a dedicated simulation program that models the behavior of the prototype detector. Hardware physicists work on the 'real' counterpart in the experiment, probing the detection capacity of a prototype with different particle beams in test beams. A comparison of simulated and test-beam data provides the basis for interpretive work, which is performed in close cooperation between the simulation expert and the corresponding hardware physicists. The epistemic gain of this twofold endeavor results from a reciprocal approximation: The models that underlie the simulation make it possible to improve the understanding of the detection mechanism; parameters are extracted from the real data and are then fitted to the simulation program that, as a result, will become 'more realistic.' Successful cooperation requires that physicists establish a common ground of understanding. This is facilitated, both socially and epistemically, by the fact that many of these software physicists have gained experience with hardware work, occasionally even 'taking shifts' and participating at the test beams.

The learning effects produced in this research environment concern both forms of expertise: simulation and hardware physics. The effects become visible when considering the combined simulation-test beam work. In the following, I shall single out a few examples of the knowledge that is produced in prototype studies and that provides significant input for the full detector simulation programs needed in later stages of the experiment.

- Geometrical features: Detectors as realized in material form will never fully match the design drawings on which their construction was based. One reason is that service parts (such as cooling pipes or wires) were given inadequate consideration in the drawings. Experimentation on a prototype allows physicists to obtain a more realistic representation of the geometrical features of the actual detector parts.
- External conditions: Testing for external conditions such as the sensitivity of the detector to temperature, humidity, or atmospheric pressure will allow physicists to understand how the signal fluctuates in response to these conditions that can

never be controlled fully in the experiment and therefore need to be accounted for numerically.

- Adaptation of heterogeneous detector parts: Prototype studies will allow physicists to better understand how the detector parts of different material interact with each other. Whereas, in first approximation, these parts are modeled in a modular way as if they were independent pieces, the prototype studies provide knowledge on how to correct for this approximation.
- Electronic noise: Every detector produces electronic noise whose origin cannot be understood in detail. Prototype studies address this problem empirically. They measure the levels of electronic noise and enable physicists to subtract it later on from the obtained signals.
- Signal shape: The shape of the signal varies from one detector to another. Knowledge about the signal shape is an important ingredient of simulation programs. It is required for reconstructing the physics processes underlying the measurements.

The acquired knowledge is included in the simulation programs in the form of new models (i.e., as new pieces of active code) or of parameters that allow one to tune and to fine-tune the different model components. The case of prototype studies evokes an alternative reading to the earlier account of simulation as constitutive of an autonomous digital laboratory. It suggests that simulation studies do not simply precede experiment. Instead, one observes a twofold dynamic:

When focusing on a particular *temporal phase*, especially the phase in which prototype studies take place, one observes the reciprocal embedding of simulation practice and hardware practice. Both feed into one another and should be considered as central elements of local practice.

When considering instead the *entire lifetime* of an experiment, the meso-level of scientific practice comes into view: Different stages of experimental practice are 'intercalated'[7] with simulation activities acting as a glue – a linkage tool that memorizes, transports, and, in a way, translates knowledge from studies of different prototypes to studies of the full detector systems. The simulation programs also incorporate and integrate knowledge that has been produced at different sites and organizational units of the collaboration, and they make it available to the collaboration as a whole. The characteristics rely on the fact that simulation work and (other forms of) experimental practice are continuously aligned and paralleled.

These very important features become visible only when considering the full spectrum of local practice as can be observed within the wider frame of the experiment. When singling out simulation as if it were to constitute a lab of its own, the second account of simulation practice in particle physics experiments remains invisible. The above observations thus raise doubt about the possibility of discussing simulation as a distinct and decoupled epistemic activity. Nicos, a physicist in the ATLAS collaboration, provides a pertinent illustration of this observation.

> The simulation, of course, you can put some things inside as you understand them but if you don't have data to compare, then the real crucial test and the criterion for physics is the data. Data decide. I mean, the data are correct. This is the statement always. Data are always correct. (MM: if you have them.) Yes, if you have them, if you have them,

> definitely. But okay, even now, ATLAS does not have the data now, but inside the simulations you definitely need to simulate the behavior of your chambers or of your silicon detectors or of your calorimeters, etc. So you need to go let's say to a test beam to understand about noise, what happens, I mean, to understand how exactly your detector is behaving. [...] You can make calculations, definitely by hand, back of the envelope calculations, but the things that finally show you a more realistic behavior of your final detector are the test beams. So you put part of your detector in the test beam before building the detector and taking the various parameters of the things you are putting finally in the Monte Carlo (Nicos, interview).

Nicos' statement that "data are always correct" introduces the 'real' apparatus as an anchor and a reference point for all experimental practice, be it simulation-based or not. Yet, his statement is programmatic as much as descriptive. He also uses it to counter a tendency to 'disembed' simulation, which not only follows from the seductive potential of simulation, but also from the extended temporal phases of the experiment, the high degree of specialization, and the highly differentiated bodies of practice.

CONCLUSIONS

This chapter has explored the association between simulation and two notions of laboratory. A *first perspective*, which emphasizes *object reconfiguration*, allows one to draw out the power and success of simulation in all phases of particle physics experiments. This success is due to the quasi-autonomous nature of simulation studies that follow their own logic throughout extended phases. The notion of 'scale reversal' (Latour 1983) accounts for the relative ease with which simulation studies are duplicated, repeated, and varied and for the reduced time scales and costs compared to the case in which the design of apparatus would have to be based on physical models (such as prototypes) and theoretical reasoning alone. In this reading, simulation is closely associated with the practice of disembedding. Reconfiguration – the transformation of objects as they occur 'in nature' into the objects worked on in laboratories – involves a form of disembedding. Reconfigured objects are easier to deal with, and it is possible to extract results from them in ways that advantage the scientist precisely *because* they have been partly disembedded from their natural environments. While this perspective is productive for analyzing important advantages of simulation, it also has its limitations. At least for the case of particle physics, the notion of a simulation laboratory in this first sense overemphasizes the potential to decouple the site of experimentation *in silico* from other sites of scientific practice.

This is where the *second perspective*, which zooms in on the broader array of locally embedded practices, provides an alternative. Instead of singling out simulation as a distinct instance of practice, it takes into consideration the complex web of intertwined practices. The prototype studies discussed above provide an example. From this angle, the disembedded practice of simulation (as sketched in the first account) appears as a mode accompanied by other modes that cover the embedding (or reembedding) tendencies of simulation work. The second perspective on the laboratory thus serves as a corrective and allows one to fill in the gaps created by an idealized account of simulation. In the case of particle physics, the scientists never forget the necessity to reembed simulated data for the purpose of the experiment. This may be explained by the fact that simulation will eventually be disciplined and needs to prove itself at the latest when real data start to flow. Consequences are, if not immediate, never far ahead. Perhaps this explains why the disembedding tendencies of simulation tend to be controlled and are kept in closely observed bounds. However, it is an open question whether the game of embedding and disembedding tendencies of simulation follows similar rhythms and patterns in other scientific domains as well. This calls for a comparative investigation of problem areas and their simulation cultures in this respect.

The material discussed here also draws attention to the importance of analyzing simulation practice according to its *temporal dynamic*, which addresses the third level of analysis mentioned in the introduction. In particle physics, the quasi-autonomous working of a digital laboratory constitutes only one phase in the life of an experiment – it is preceded, framed, paralleled, and succeeded by other phases. This observation suggests that one should work with a combined framework that renders fruitful both perspectives on the laboratory: to consider simulation intermittently as following a logic of reconfiguration and a logic of locally intertwined heterogeneous practice. This includes an attention to the disembedding *as well as* the (re-)embedding movements of object work. Applying a combined framework of this sort to the study of simulation practice allows one to unravel the multilayered structure that simulation displays and that accounts for its power and versatility. Particle physics is a particularly interesting case because it exploits simulation's rich potential in all its dimensions.

Finally, the observation that attention should also be given to the reembedding movements of object work raises a more general issue. One might actually wonder whether the laboratory studies of STS have dealt sufficiently with the question of how laboratory results become transferred across the boundaries of the lab into other areas of science or society (see Merz 2006). This involves questions such as the one about the strategies that ensure that results elaborated in a laboratory can be transferred successfully (as concerns their epistemic and social dimensions) to other realms. In the case of simulation in particle physics, managing the applicability of its results to the conduct of experiments is of utmost importance. While particle physicists in general will agree with this diagnosis, a separate culture of simulation is developing that might, at times, lose this insight from view.

ACKNOWLEDGMENTS

For helpful comments I thank Erika Mattila, Martin Meister, and the volume's three editors. I am grateful to the physicists for their patience and interest.

University of Lausanne, Switzerland, and Eidgenössische Materialprüfungs- und Forschungsanstalt, EMPA, St. Gallen, Switzerland

NOTES

[1] A loosely related perspective on modeling and simulation addresses the issue of 'mediation,' considering models and simulations as intermediaries between theory and experiment and analyzing their shifting positions in this spectrum (see Galison 1997; Morgan and Morrison 1999; Winsberg 1999).

[2] For a thoughtful account challenging Latour's claim that laboratories (in all cases) "raise the world," see Scott (1991).

[3] The material presented in this chapter stems from participant observation of one of the LHC collaborations in a preparatory phase in which the detector was being designed and constructed. Insight into the role of simulation in other experimental phases was obtained by interviewing physicists working on other experiments such as those at the electron-positron collider LEP. The observed scientists include theorists, experimentalists, and computer scientists – all of whom are involved with simulation work at CERN.

[4] This level of analysis, although pervading the chapter, has to remain implicit for lack of space. It concerns the power of simulation to integrate heterogeneous practices within a collaboration and throughout different phases of an experiment (for a first analysis, see Merz 2005).

[5] The scientists' names have been replaced with pseudonyms.

[6] In fact, the different phases are not as neatly separated as may appear. Studies often start before the first phase has been concluded, resulting in a partial overlap of consecutive experimental phases.

[7] Galison (1997: chap. 9) introduces the concept of intercalation to describe a slightly different dynamics of theory-experiment-instrumentation relations.

REFERENCES

Dowling, D. (1999). "Experimenting on theories", *Science in Context*, **12** (2): 261–273.

Galison, P. (1997). *Image and Logic. A Material Culture of Microphysics*, Chicago, IL: University of Chicago Press.

Keller, E.F. (2003). "Models, simulation, and 'computer experiments'", in H. Radder (ed.), *The Philosophy of Scientific Experimentation*, Pittsburgh, PA: University of Pittsburgh Press, pp. 198–215.

Knorr-Cetina, K. (1992). "The couch, the cathedral, and the laboratory: On the relationship between experiment and laboratory in science", in A. Pickering (ed.), *Science as Practice and Culture*, Chicago, IL: University of Chicago Press, pp. 113–137.

Knorr-Cetina, K. (1995). "How superorganisms change: Consensus formation and the social ontology of high-energy physics experiments", *Social Studies of Science*, **25**: 119–147.

Knorr-Cetina, K. (1999). *Epistemic Cultures. How the Sciences Make Knowledge*, Cambridge, MA: Harvard University Press.

Latour, B. (1983). "Give me a laboratory and I will raise the world", in K. Knorr-Cetina and M. Mulkay (eds.), *Science Observed*, London: Sage, pp. 141–170.

Latour, B. and S. Woolgar (1979). *Laboratory Life: The Social Construction of Scientific Facts*, Beverly Hills, CA: Sage.

Lynch, M. (1997). *Scientific Practice and Ordinary Action: Ethnomethodology and Social Studies of Science*, Cambridge, UK: Cambridge University Press.

Merz, M. (1999). "Multiplex and unfolding: Computer simulation in particle physics", *Science in Context*, **12** (2): 293–316.

Merz, M. (2002). "Kontrolle – Widerstand – Ermächtigung: Wie Simulationssoftware Physiker konfiguriert", in W. Rammert and I. Schulz-Schaeffer (eds.), *Können Maschinen handeln? Soziologische Beiträge zum Verhältnis von Mensch und* Technik, Frankfurt a. M.: Campus, pp. 267–290.

Merz, M. (2005). "The performative idiom of computer simulation", manuscript presented at the Workshop *A New Form of Experiment? Exploring the Performative Potential of Models and Computer Simulations*, Claremont, CA: Harvey Mudd College.

Merz, M. (2006). "The topicality of the difference thesis: Revisiting constructivism and the laboratory", Special issue *STS after constructivism, Science, Technology & Innovation Studies*, forthcoming.

Morgan, M.S. (2003). "Experiments without material intervention: Model experiments, virtual experiments, and virtually experiments", in H. Radder (ed.), *The Philosophy of Scientific Experimentation*, Pittsburgh, PA: University of Pittsburgh Press, pp. 216–235.
Morgan, M.S. and M. Morrison (eds.) (1999), *Models as Mediators: Perspectives on Natural and Social Science*, Cambridge, UK: Cambridge University Press.
Scott, P. (1991). "Levers and counterweights: A laboratory that failed to raise the world", *Social Studies of Science*, **21** (1): 7–35.
Sismondo, S. (1999). "Models, simulations, and their objects", *Science in Context*, **12** (2): 247–260.
Winsberg, E. (1999) "Sanctioning models: The epistemology of simulation", *Science in Context*, **12** (2): 275–292.

CHAPTER 11

ARTHUR C. PETERSEN

SIMULATION UNCERTAINTY AND THE CHALLENGE OF POSTNORMAL SCIENCE

INTRODUCTION

On January 20, 1999, Hans de Kwaadsteniet, a senior statistician at the Netherlands National Institute for Public Health and the Environment (*Rijksinstituut voor Volksgezondheid en Milieu*, RIVM), made news in the Netherlands. After years of trying to convince his superiors that the environmental assessment branch[1] of the institute leaned too much toward computer simulation at the expense of measurements, he went public with this criticism by publishing an article on the op-ed page of the national newspaper *Trouw* (de Kwaadsteniet 1999). His article was supplemented with an interview that resulted in the headline "Environmental Institute Lies and Deceives" on the newspaper's front page. His specific claim was that the RIVM was suggesting an excessive accuracy for environmental figures published yearly in its *State of the Environment* report. According to him, too many model results were included that had not been compared rigorously with observational data – mostly because of the lack of sufficiently detailed data to do the necessary comparisons. He pointed out that living in an "imaginary world" was dangerous. He thought that if the institute spent more time and energy on testing and developing computer-simulation models in a way that were to make more use of existing and newly performed observations, it would become more careful in the way it presented its results to policy makers. De Kwaadsteniet identified the deceiving speed, clarity, and internal consistency of the computer-simulation approach as the main causes of the claimed bias toward computer simulation at RIVM.

The institute responded immediately to the publication by suspending de Kwaadsteniet from his job and stating in an official reaction that a significant fraction of its environmental research budget was spent on observations, that no policy recommendations were given when uncertainties were too large, and that the uncertainties were not left out of the *State of the Environment* reports on purpose. The institute promised to publish information on the uncertainties in next editions. In a later reaction, the institute's Director of the Environment, Klaas van Egmond (1999), argued that simulation models must be viewed as "condensed knowledge" and that they are

J. Lenhard, G. Küppers, and T. Shinn (eds.), Simulation: Pragmatic Construction of Reality, 173–185.

indispensable in environmental assessment, since, without them, it would be impossible to determine cause-effect relationships between sources and effects of pollution. Thus, models give meaning to measurement results. And they are needed in environmental policy making. Furthermore, he observed that policy makers often are confronted with incomplete knowledge, and that the institute regards it as its task to report on the current state of affairs in the environment, including the uncertainties involved. He gave the example that it will take many more years before climate research reaches the 'ultimate truth' about what is happening to the climate. However, on the basis of currently available knowledge and its uncertainties, politicians have to decide whether to take measures now already. Finally, the Director added that the most important conclusions contained in the summaries for policy makers of the *State of the Environment* reports are carefully crafted, taking all relevant uncertainties into account.

Soon after the publication by de Kwaadsteniet, an intense and long-lasting media debate ensued in the Netherlands.[2] The affair reached the floor of the Dutch Parliament within a matter of days. Facing Parliament, the Minister of the Environment, Jan Pronk, defended the integrity of the institute. In return for an agreement to organize more regular external reviews of its environmental assessment activities and improve its communication of uncertainty, the Minister granted the institute additional funding for its monitoring activities.

The episode of de Kwaadsteniet's questioning of the role of scientific simulation in politics is by no means unique in the world. Controversies like that in the Netherlands surface regularly in many countries. In such discussions, general questions arise about the role of simulation in science as well as its role in policy making. The latter question constitutes the subject of this chapter. It focuses particularly on the reliability of simulation for political uses.

Since World War II, computational approaches in science have emerged and expanded – not in isolation, but often in strong contact with experimental and observational fields in the natural sciences, and aided by developments in mathematics and computer science.[3] Outside science as well, simulations have become important tools in, for instance, providing scientific advice to policy makers.[4] In highly politicized cases, such as climate change, methodological questions about what constitutes 'good' or 'sound' science, often left implicit in scientific practices, are brought into the open. The characteristics of 'sound' science on which policies can be based are contested in political forums. Typically, the issue of the reliability of computer simulation plays an important role in these debates. The state of affairs in which there are high political stakes in conjunction with high systems uncertainty has given rise to normative appeals for systematically dealing with uncertainty in scientific policy advicing. Funtowicz and Ravetz's (1991) proposal for a "post-normal science" problem-solving strategy constitutes a prominent example.

This chapter first discusses the use of scientific simulation for policy. Subsequently, after treating general issues related to the science–policy interface and the challenge of postnormal science, it presents a case study on simulating climate change along with a new methodology for assessing and communicating uncertainty in science-for-policy developed by RIVM and external uncertainty experts in response to the media affair. Finally, it outlines the implications of this new methodology for managing the use of simulation in science-for-policy.

SIMULATION UNCERTAINTY IN SCIENCE-FOR-POLICY

Scientific simulation models are not just used within science. The results of scientific simulation models are frequently employed in public policy making as well.[5] How important it is to assess computer-simulation uncertainty, as part of the process of providing scientific advice to policy makers and politicians, depends on the questions asked of science. In cases in which policy makers ask questions "which cannot be answered by science" (Weinberg 1972) – that is, even though the questions are scientifically formulated, the uncertainties are too large to answer those questions unequivocally – it is typical that many different answers can be produced by applying different simulation models to the policy issue. In this chapter, it is argued that, in such cases, computer-simulation uncertainty must be assessed thoroughly – and that this must be done in a way that is appropriate to the decision-making context.

Similar arguments were developed by the philosophers of science Silvio Funtowicz and Jerry Ravetz (e.g., 1990, 1991, 1993). The aim of Funtowicz and Ravetz's work is to improve the decision-making process by introducing into the policy-advisory process appropriate information about the uncertainty and quality of the underlying science ("science providing advice to policy" can be called "science-for-policy" in short). In the Prologue to their book, this aim is set in the following context:

> There is a long tradition in public affairs which assumes that solutions to policy issues should, and can, be determined by 'the facts' expressed in quantitative form. But such quantitative information, either as particular inputs to decision-making or as general purpose statistics, is itself becoming increasingly problematic and afflicted by severe uncertainty. Previously it was assumed that Science provided 'hard facts' in numerical form, in contrast to the 'soft', interest-driven, value-laden determinants of politics. Now, policy makers increasingly need to make 'hard' decisions, choosing between conflicting options, using scientific information that is irremediably 'soft' (Funtowicz and Ravetz 1990: 1).

In Funtowicz and Ravetz's analysis, the stated "softness" of the scientific information relates mainly to their claim that for many pressing policy problems, we cannot draw on the reliable knowledge that can be gained from experiments, but instead must use much less reliable knowledge from simulation. Even though one cannot make the general statement that all simulations are less reliable than laboratory experiments, nor that simulations are in all respects untestable, the question of the reliability of simulation is indeed pressing for the particular cases discussed by Funtowicz and Ravetz – that is, very complicated and complex environmental issues.

Especially in simulation studies of the future, we must recognize our ignorance about the complex systems under study. Verification and validation of these computer models is impossible, and confirmation is inherently partial. Furthermore, since models are products made by scientists, we must always be aware of the possible presence of personal, institutional, or ideological dimensions – their potential 'value-ladenness.' Knowledge claims based on simulation should be tailored to be insensitive to artifactual aspects of models and precise about real effects (Norton and Suppe

2001: 84). In order to be able to tailor these claims to such requirements, simulationists must, on the one hand, do much practical work to determine the sensitivity of their model results to all sources of uncertainty. It is often not feasible, however, to establish the reliability of a simulation in quantitative terms. Therefore, one has to also assess the reliability of a model in a qualitative manner, for which a thorough review of the model is usually helpful. But even then, the quality of the simulation is only established according to the scientific community's methodological standards. Finally, the scientists must think about remaining uncertainties that have not been estimated (yet) and determine what they can say about them. All these steps require a substantial amount of work. Since policy makers are usually not able to judge the reliability of scientific simulation-model outcomes themselves, scientific policy advisers must carefully assess the reliability of their simulations and be aware of the uncertainties in the presentation of their conclusions.[6]

Simulation models of ecological systems, for example, although they may give an impression of the scope of behavioral possibilities of such systems and, as such, may contribute reasons for taking policy measures, cannot predict the future states of these open and unpredictable systems. If modeling assumptions were made in a more transparent manner, and if, in concrete problem contexts, all relevant policy actors were involved in the framing of the models (what questions to address, where to locate the system boundaries, etc.), the choice of the models, and the evaluation of the models, then

> [m]odelling could ... contribute to the organization of knowledge, e.g. it could catalyze mutual learning processes and it could contribute to the integration of scientific and non-scientific knowledge and of exo- en endo-perspectives [perspectives from respectively outside or within the system studied] (Haag and Kaupenjohann 2001: 57).

This is proposed as an ideal situation. Current practice is far from this ideal, however. Leaving aside the question of whether the ideal can ever be reached, we can observe that simulation uncertainties do not often get the airing they may well deserve. Sometimes, policy makers, politicians, and other actors do not see a need to dwell on the uncertainties and treat them explicitly. Policy decisions are just taken without being explicit about the level of uncertainty of the risk involved. A concrete example from the area of international environmental policy making is the formation of the *Mediterranean Action Plan* (Med Plan), a regional environmental cooperation for dealing with the issue of marine pollution in the Mediterranean that arose in the 1970s. The uncertainty in this example is related to uncertainties in ecotoxicological simulations. The main scientists and policy makers involved in the Med Plan "shared an abiding belief in ecological principles and were committed to preserving the physical environment, which they thought was threatened by pollution" (Haas 1990: 74–75). These ecological principles were partly derived from theoretical ecological computer simulations used to study the behavior of complex ecological systems. These simulations are relatively unreliable. This did not seem to hinder the main policy actors. The uncertainties in ecological computer simulations were dealt with only implicitly, not explicitly, by the actors involved in the Med Plan and remained at an unreflective level while decisions were being taken. Increased transparency about simulation uncertainties need not have changed the same policy outcomes, but would have made the decisions more robust against these uncertainties. An explicit precautionary approach could have been used, for instance.

Recently, however, national and international governmental bodies have undergone a reflective transition in their attitudes toward scientific uncertainty. At the end of the 1980s, when environmental policy makers were faced with significant scientific uncertainties surrounding large-scale and high-impact environmental problems such as biodiversity loss and climate change, they started referring more and more often to the "precautionary principle," for instance. Loosely formulated, the principle states that if there is evidence that a certain activity may be harmful to humans or the environment, that activity should be abandoned. The principle provides politicians with the possibility to install measures even when uncertainty still exists about a problem.[7]

Thus, scientific simulation models are often used in providing policy advice, and they typically have significant uncertainties attached to them. In practice, it turns out that many experts still find it difficult to deal with these uncertainties when providing their policy advice. Within their own disciplines, they typically do not learn the skills needed to deal adequately with these uncertainties when providing advice (van Asselt and Petersen 2003: 144–145). There is clearly a need for including these issues in core academic curricula.

THE CHALLENGE OF POSTNORMAL SCIENCE

Many social scientists who have studied the relationship between science and decision making have concluded that these two activities cannot be separated neatly in practice. One way to phrase this conclusion is the following: "Natural knowledge and political order are co-produced through a common social project that shores up the legitimacy of each" (Jasanoff and Wynne 1998: 16). An example may serve to illustrate this point.

A much-discussed, though exceptional, coproduction of natural knowledge and political order is the ongoing assessment process conducted by the Intergovernmental Panel on Climate Change (IPCC), which receives questions from and feeds back into the United Nations Framework Convention on Climate Change. Due to widely publicized warnings from scientists in the 1980s, the public in Western democracies became interested in the risks involved in an enhanced greenhouse effect induced by anthropogenic emissions of CO_2 leading to a human-induced global warming – and its associated effects, such as sea-level rise. The attribution of climate change to human influences and the projections of climate change into the future have made heavy use of climate simulations. Since the societal changes implied by the different solutions proposed for solving the global warming problem are quite drastic, one of the first steps politicians took to address the problem was to ask scientists to regularly assess the state of climate science as well as the possibilities for adaptation to climate change and mitigation of the problem by reducing anthropogenic greenhouse gas (mostly CO_2) emissions. This led to the establishment of the IPCC in 1988.[8] The advisory process involving the IPCC is regarded by many social scientists as being a 'co-production' of, on the one hand, our knowledge about the climate system and, on the other hand, the international political order:

> The IPCC's efforts to provide usable knowledge resonated with the belief of sponsoring policy organizations that climate change is a manageable problem within the framework of existing institutions and cultures (Jasanoff and Wynne 1998: 37).

The alignment of scientific and political views seems to be a common feature of environmental assessment (see, e.g., Haas 1990 for a similar analysis of science and policy involved in the Med Plan). From these and other examples, one may conclude that the knowledge used in scientific assessments for policy purposes, often largely based on computer simulations, is potentially 'value-laden.'

Already earlier in the 1980s, before the IPCC was established, the special challenges facing experts under conditions of potential alignment of scientific and political views became evident in the area of risk assessment. Recognizing that the interactions between science and policy making on risks were often unproductive in cases in which the decision stakes and system uncertainty are very high, Funtowicz and Ravetz proposed to distinguish a new type of risk assessment called "total-environmental assessment" (Funtowicz and Ravetz 1985: 228). This is a form of risk assessment in which the "total environment" – that is, the complete context – of a risk issue is taken into account. This kind of risk assessment is appropriate for cases with high decision stakes and system uncertainty.[9] In very polarized settings, the least one can hope for, according to Funtowicz and Ravetz (1985: 229), is a "consensus over salient areas of debate."

According to Funtowicz and Ravetz, structural changes in the direction of enhanced participation are needed in order to democratize scientific advisory proceedings. For this reason, they have generalized their original normative view on risk assessment into a sweeping normative statement on the future of science-for-policy:

> Now global environmental issues present new tasks for science; instead of discovery and application of facts, the new fundamental achievements for science must be in meeting these challenges. ... In this essay, we make the first articulation of a new scientific method, which does not pretend to be either value-free or ethically neutral. The product of such a method, applied to this new enterprise, is what we call 'post-normal science' (Funtowicz and Ravetz 1991: 138).

When Funtowicz and Ravetz first wrote about "risk assessment," they subsequently generalized their analysis to "problem-solving strategies." The problem-solving strategy of "postnormal science" (or 'second-order science') corresponds to the "total-environmental" type of risk assessment discussed above (Funtowicz and Ravetz 1991: 137, 144–145).[10]

Whether or not one agrees with Functowicz and Ravetz's statement that "science" as a whole has to tackle the "new tasks," whoever takes up the challenge has the responsibility to conscientiously (a) assess the issues, which may involve building very complicated computer simulations; (b) assess the uncertainties; and (c) communicate the policy-relevant findings of both these assessment activities.

ASSESSING AND COMMUNICATING SIMULATION UNCERTAINTY IN SCIENCE-FOR-POLICY

How should we deal with the challenge that postnormal science poses to the use of computer simulation in policy making? Let us take a look at a specific example, that of climate simulation and climate policy. Climate simulations play an important role in climate science. These simulations involve mathematical models that are implemented on computers and imitate processes in the climate system. Like the history of numerical weather prediction, the history of climate science is strongly related to the history of the computer. There are two main reasons why simulation is so important in climate science. First, computers removed an actual barrier in meteorological practice: They greatly enhanced the speed with which calculations could be done. The calculations in climate simulations cannot be done practically without the use of computers. Second, simulation is an important ingredient of climate science, because real experiments with the climate as a whole are impossible. If we want to manipulate climate 'experimentally,' we need to perform such manipulations on a digital representation of the climate system.

It must be borne in mind here that climate science is an observational science in which the scientific activities encompass much more than performing computer simulations. In fact, climate observations are of pivotal importance – also for climate-simulation practice. From climate observations, the world's climate scientists have concluded that it is very likely that the earth's climate has changed over the last 100 years. In 2001, the *Third Assessment Report of the Intergovernmental Panel on Climate Change* (IPCC) concluded that the global average surface temperature has increased by 0.6 ± 0.2°C (95% confidence range) over this period (IPCC 2001: 2). The uncertainty is expressed here as a range of temperature change (from 0.4 to 0.8°C) together with the probability that the real value lies within this range (that is, 95%). For the Northern Hemisphere, it is considered likely (a judgmental estimate of confidence that there is a 66 to 90% chance) that current temperatures are higher than historic temperatures over the last millennium (IPCC 2001: 2).

Alongside temperature, precipitation is also a component of climate. It is considered very likely that precipitation has increased by 5 to 10% during the twentieth century over most mid- and high latitudes of the Northern Hemisphere continents (IPCC 2001: 4). Furthermore, in the mid- and high latitudes of the Northern Hemisphere it is likely, according to the climate experts, that there has been a 2 to 4% increase in the frequency of heavy precipitation events over the latter half of the twentieth century (IPCC 2001: 4). Also such extreme events are typically included in the description of climate.

The above statements about observed climate change have been obtained without the use of climate simulations. This means that the sources of uncertainty are of a different kind to those encountered in simulation practice. For example, for global average surface temperature, the sources of uncertainty on the 100-year timescale are located in data and (statistical) model assumptions made in data processing: "data gaps, random instrumental errors and uncertainties, uncertainties in bias corrections in the ocean surface temperature data and also in adjustments for urbanisation over the land" (IPCC 2001: 3). For the Northern Hemisphere temperature on the 1,000-year timescale, the sparseness of 'proxy' data[11] is the main source of uncertainty (IPCC 2001: 3), besides the unreliability of proxies for determining local temperatures in the past.

It is not possible, however, to deduce the *causes* of the observed changes in climate directly from the observations. When climate scientists want to attribute climate changes to causes or make future projections, they need to make use of climate simulations. One of the most important conclusions of the IPCC (2001) is that "most of the observed warming over the last 50 years is likely [between 66 and 90% chance] to have been due to the increase in greenhouse gas concentrations" (IPCC 2001: 10). In order to arrive at this conclusion, climate simulations have been performed as a substitute for experiments. This function of simulation is crucial in climate science, because there is only one historical manifestation of the system under study. Real (in the sense of controlled and reproducible) experiments on the scale of the whole climate system are impossible.

The roles of climate simulation in climate science are manifold. Furthermore, climate models of varying levels of concreteness exist and are valued differently by different groups of climate scientists. On the one hand, we find relatively simple climate models that do not require huge computational resources but can be used for genuine climate-scientific research. On the other hand, we encounter very comprehensive climate models that demand high-end supercomputers in order to be able to work with them. For this latter category of climate models, computing power is currently a bottleneck. This situation will remain unchanged for at least the next decade (the demand for computational power will keep growing faster than what can be delivered). The IPCC reports have taken a pragmatic stance in this matter and acknowledge that both comprehensive and simple models have important roles to play in climate science (see, also, Petersen 2000). The observed plurality at the methodological level is correlated with a plurality of aims and goals held by climate-simulation practitioners in their scientific practice. The social context of climate-simulation practice has a significant influence on this practice. Thus, in evaluating climate simulations, the potential value-ladenness of choices should not be overlooked.

Even though all climate models contain ad hoc 'parameterizations' and can be criticized methodologically for that reason, climate scientists generally feel confident about using these models for climate-change studies. However, the IPCC lacks a methodology for uncertainty assessment and a typology of uncertainty that can be used to assess uncertainties more systematically. The challenge to postnormal science is for the IPCC to become even more rigorous and transparent in its treatment of uncertainty.

The MNP faces a similar challenge. In the year 2000, the MNP identified the lack of systematic treatment of uncertainty in the area of environmental policy making as one of the causes of the media affair reported at the beginning of this chapter. In order to help environmental assessors to deal with uncertainty and frame policy problems in a more appropriate way, the Netherlands Environmental Assessment Agency (*Milieu- en Natuurplanbureau*, MNP), then part of RIVM, together with Utrecht University and an international team of uncertainty experts, developed the *RIVM/MNP Guidance for Uncertainty Assessment and Communication* (Petersen et al. 2003; Janssen et al. 2003; van der Sluijs et al. 2003; van der Sluijs et al. 2004).

The *RIVM/MNP Guidance for Uncertainty Assessment and Communication* (www.mnp.nl/guidance) offers assistance to employees of the Netherlands Environmental Assessment Agency in mapping and communicating uncertainties in environmental assessments.[12] It was judged that the Guidance should facilitate dealing with uncertainties throughout the whole environmental assessment process and not be limited to applying ready-made tools for uncertainty analysis and communication, because choices are made in all parts of environmental assessments that influence the way uncertainties are dealt with. The way in which the perspectives of other scientists and stakeholders are treated is particularly crucial when assessing relatively unstructured policy problems.

The Guidance identifies six parts of environmental assessments that have an impact on the way uncertainties are dealt with. These parts are:

1. problem framing;
2. involvement of stakeholders (i.e., all those involved in or affected by a policy problem including experts);
3. selection of indicators representing the policy problem;
4. appraisal of the knowledge base;
5. mapping and assessment of relevant uncertainties;
6. reporting of the uncertainty information.

A focused effort to analyze and communicate uncertainty is usually made in parts 5 and 6. However, the choices and judgments made in the other four parts are also of high importance for dealing with uncertainty.

The Guidance is not set up as a protocol. Instead, it aspires to stimulate reflection on the choices made in different parts of environmental assessments, in order to make them more conscious and produce a better way of dealing with uncertainties. Aside from stimulating reflection during the execution of environmental assessments, the Guidance is intended to signal in a timely way which bottlenecks might occur when dealing with uncertainties (and what additional effort should perhaps be made in the field of uncertainty assessment). The Guidance offers advice on the selection of methods and tools for adequately estimating uncertainties in the given context and communicating them to scientific researchers, the 'clients' (usually ministries), other actors in the policy process, and the broader public. The group of envisaged users of the Guidance comprises a large fraction of the employees of the Netherlands Environmental Assessment Agency (among others, those who fulfill the roles of project leader, project-team member, researcher, or policy adviser).

The Guidance can be used in different phases of a project (at the beginning, during, after). *At the beginning* of a project, it can play an important role in designing and elaborating the way uncertainty will be dealt with during the project. *During* a project, the Guidance can be of assistance in performing the uncertainty assessment and communicating the results. *After* a project, it can be of use in reviewing and evaluating the project.

The most important function of the instrument is to make the practitioners reflect on the importance of uncertainties and on the way they should communicate these uncertainties to stakeholders (including policy makers). Table 1 shows the uncertainty typology used in the Guidance.[13] The Guidance typology is presented as a matrix. This 'uncertainty matrix' is based on five dimensions of uncertainty. In the Guidance, it is used as an instrument for generating an overview of where one expects the most important (policy-relevant) uncertainties to be located (the first dimension), and how these can be further characterized in terms of four other uncertainty dimensions.

Table 1. Uncertainty matrix[14]

Location of uncertainty ↓		Level of uncertainty (*from determinism, through probability and possibility, to ignorance*)			Nature of uncertainty		Qualification of knowledge base			Value-ladenness of choices		
		Statistical uncertainty	Scenario uncertainty	Recognized ignorance	Epistemic	Variability	–	0	+	–	0	+
Context												
Expert judgment												
MODEL	Structure											
	Implementation											
	Parameters											
	Inputs											
Data												
Outputs												

Using the matrix can serve as a first step toward a more elaborate uncertainty assessment in which the size of uncertainties and their impact on the policy-relevant conclusions is assessed explicitly.[15] For further details about the Guidance, the reader is referred to the Guidance website and publications.

This typology of simulation uncertainty can be applied fruitfully in the analysis of climate-simulation uncertainty, as is shown for the simulation-related sources of uncertainty in climate-change attribution studies by Petersen (in preparation). By applying the typology, it becomes immediately obvious that only part of the uncertainty can be expressed statistically. Additional qualitative judgments on the methodological quality of the climate-simulation models (qualification of the knowledge base) are needed – and indeed played an important role in the production of the IPCC (2001) report. Since the vocabulary needed to explicitly distinguish between the two uncertainty sorts of statistical uncertainty ("inexactness" in the vocabulary of Funtowicz and Ravetz 1990) and qualification of the knowledge base (methodological "unreliability" according to Funtowicz and Ravetz 1990) was not available to the lead authors, the influence of their qualitative judgments on reaching their final conclusion remained largely invisible to outsiders.

Since the Guidance was released in December 2002, it has become part of the agency's system of quality assurance for all projects including those making heavy use of simulations. Through teaching courses, an increasing proportion of scientific advisers have become acquainted with the new methodology. Specific tools for uncertainty assessment that are presented in the Guidance have demonstrated their usefulness for prioritizing research activities in simulation modeling, for instance, the research on the global energy simulation model TIMER (van der Sluijs et al. 2002).

Even though the Guidance is only an instrument for reflection, a change in simulation practice can be observed within the agency in the sense that modeling choices are made more reflectively and reports pay more attention to uncertainties. It remains to be seen whether the institute has become less vulnerable to media affairs such as the one caused by de Kwaadsteniet, but my contention is that the answer will be positive.

CONCLUSION

The facts that science and policy cannot be separated neatly and that experts provide policy advice under conditions of high political stakes and high system uncertainty pose a severe challenge to those expert advisers who use scientific simulation models that have significant uncertainties attached to them. From their own disciplines, experts typically do not gain the necessary skills to adequately deal with these uncertainties when providing their advice. By making systematic use of an instrument such as the RIVM/MNP Guidance on Uncertainty Assessment and Communication, experts are better able to meet the challenge of the postnormal science problem-solving strategy.

Netherlands Environmental Assessment Agency (MNP), Bilthoven, The Netherlands

NOTES

[1] Over the years, this branch has become an independent part of the RIVM: The Netherlands Environmental Assessment Agency (*Milieu- en Natuurplanureau*, MNP).

[2] See, for more information about this debate, van Asselt (2000) and van der Sluijs (2002).

[3] Computer simulation as a scientific approach is not limited to the natural sciences, however. Simulation is gaining ever more prominence in, for example, psychology, sociology, political science, and economics. The recent rise in the amount of work on simulation in these fields may be partly related to the wide applicability of the concept of 'complex systems' (see Casti 1997, who provides a popularized account of the use of simulation to study complex systems in the natural and social sciences). Many simulations in both the natural and social sciences share system-theoretical concepts.

[4] Other examples of the use of simulation techniques outside science are flight simulators for training pilots and simulations used in technology development as tools to design and 'test' new technologies, be they in automobile design (simulations of aerodynamics or crashes) or nuclear weapons design (simulation of stockpile safety or explosions).

[5] Not all scientific simulation models find their application in policy making. This chapter only deals with those models that do.

[6] Obviously, there is also a more general need to provide insight into the uncertainties involved in policy advice, and not just in the case of scientific computer simulation. Whereas the main emphasis of this chapter is on simulation-model uncertainty, the general discussion on the science–policy interface and assessing uncertainty in science-for-policy does not just apply to scientific simulation.

[7] Many references can be given to literature on the precautionary principle. Petersen and van der Zwaan (2003) offer a concise introduction to the principle and how it relates to the responsibility of scientific advisers to communicate about uncertainties.

[8] The IPCC consists of three working groups. Currently, Working Group I deals with the (natural) scientific basis of climate change; Working Group II addresses issues of impacts, adaptation, and vulnerability; and Working Group III assesses mitigation options. The analysis presented in this chapter focuses on Working Group I.

[9] These two variables are not totally independent, in the sense that the recognition of system uncertainty is typically enhanced if the decision stakes are high (see Jasanoff and Wynne 1998: 12).

[10] The other two types of problem-solving strategies are applied science (low systems uncertainty and/or low decision stakes) and professional consultancy (medium-level systems uncertainty and/or medium-level decision stakes) (e.g., Funtowicz and Ravetz 1991, 1993).

[11] 'Proxies' such as tree rings, corals, ice cores, and historical records are "interpreted, using physical and biophysical principles, to represent some combination of climate-related variations back in time" (IPCC 2001: 795).

[12] Only some elements of the Guidance are specific to environmental assessment, however. With only some minor changes, the Guidance can be used in any science-for-policy activity. Furthermore, although a strong emphasis is placed on assessing simulation uncertainty, the methodology encompasses all sources of information used in science-for-policy.

[13] This uncertainty typology is based partly on a paper by Walker et al. (2003). That paper was the result of a process involving some of the uncertainty experts who also participated in developing the *Guidance*. In Walker et al. (2003), uncertainty is classified according to three dimensions: its 'location' (where it occurs), its 'level' (where uncertainty manifests itself on the gradual spectrum between deterministic knowledge and total ignorance), and its 'nature' (whether uncertainty primarily stems from knowledge imperfection or is a direct consequence of inherent variability). Janssen et al. (2003) have extended this typology by adding two additional dimensions (represented by two columns on the right-hand side of the uncertainty matrix) denoted 'qualification of knowledge base' and 'value-ladenness of choices.' In order to make the uncertainty matrix more widely applicable than in model-based decision support studies, two location categories have been added, namely 'expert judgment' and 'data.'

[14] Table adapted from Janssen et al., 2003, with permission from *The Netherlands Environmental Assessment Agency, National Institute for Public Health and the Environment RIVM.*

[15] This is done by directly linking the different cells in the matrix to a list of uncertainty-assessment tools (van der Sluijs et al. 2004).

REFERENCES

Casti, J.L. (1997). *Would-Be Worlds: How Simulation is Changing the Frontiers of Science*, New York: John Wiley & Sons.

de Kwaadsteniet, H. (1999). "De samenleving heeft recht op eerlijke informatie, het RIVM geeft die niet" ("Society has a right to honest information, which RIVM does not provide"), *Trouw*, January 20: 16.

Funtowicz, S.O. and J.R. Ravetz (1985). "Three types of risk assessment: A methodological analysis", in C. Whipple and V.T. Covello (eds.), *Risk Analysis in the Private Sector*, New York: Plenum Press, pp. 217–231.

Funtowicz, S.O. and J.R. Ravetz (1990). *Uncertainty and Quality in Science for Policy*. Dordrecht, NL: Kluwer Academic Publishers.

Funtowicz, S.O. and J.R. Ravetz (1991). "A new scientific methodology for global environmental issues", in R. Constanza (ed.), *Ecological Economics: The Science and Management of Sustainability*, New York: Columbia University Press, pp. 137–152.

Funtowicz, S.O. and J.R. Ravetz (1993). "Science for the post-normal age", *Futures*, **25**: 739–755.

Haag, D. and M. Kaupenjohann (2001). "Parameters, prediction, post-normal science and the precautionary principle: A roadmap for modelling for decision-making", *Ecological Modelling*, **144**: 45–60.

Haas, P.M. (1990). *Saving the Mediterranean: The Politics of International Environmental Cooperation*, New York: Columbia University Press.

IPCC (2001). *Climate Change 2001: The Scientific Basis. Contribution of Working Group I to the Third Assessment Report of the Intergovernmental Panel on Climate Change*, Cambridge, England: Cambridge University Press.

Janssen, P.H.M., A.C. Petersen, J.P. van der Sluijs, J.S. Risbey, and J.R. Ravetz (2003). *RIVM/MNP Guidance for Uncertainty Assessment and Communication: Quickscan Hints & Actions List*, Bilthoven, NL: Netherlands Environmental Assessment Agency (MNP), National Institute for Public Health and the Environment. Available at:
http://www.rivm.nl/bibliotheek/digitaaldepot/Guidance_QS-HA.pdf.

Jasanoff, S. and B. Wynne (1998). "Science and decisionmaking", in S. Rayner and E.L. Malone (eds.), *Human Choice and Climate Change, Vol. 1: The Societal Framework*, Columbus, OH: Batelle Press, pp. 1–87.

Norton, S.D. and F. Suppe (2001). "Why atmospheric modeling is good science", in C A. Miller and P.N. Edwards (eds.), *Changing the Atmosphere: Expert Knowledge and Environmental Governance*, Cambridge, MA: MIT Press, pp. 67–105.

Petersen, A.C. (2000). "Philosophy of climate science", *Bulletin of the American Meteorological Society*, **81**: 265–271.

Petersen, A.C. (in preparation). *Simulating Nature: A Philosophical Study of Computer-Simulation Uncertainties and Their Role in Climate Science and Policy Advice.*

Petersen, A.C. and B.C.C. van der Zwaan (2003). "The precautionary principle: (Un)certainties about species loss", in B.C.C. van der Zwaan and A.C. Petersen (eds.), *Sharing the Planet: Population – Consumption – Species. Science and Ethics for a Sustainable and Equitable World*, Delft, NL: Eburon Academic Publishers, pp. 133–150.

Petersen, A.C., P.H.M. Janssen, J.P. van der Sluijs, J.S. Risbey, and J.R. Ravetz (2003). *RIVM/MNP Guidance for Uncertainty Assessment and Communication: Mini-Checklist & Quickscan Questionnaire*. Bilthoven, NL: Netherlands Environmental Assessment Agency (MNP), National Institute for Public Health and the Environment. Available at:
http://www.rivm.nl/bibliotheek/digitaaldepot/Guidance_MC_QS-Q.pdf.

van Asselt, M.B.A. (2000). *Perspectives on Uncertainty and Risk: The PRIMA Approach to Decision Support*, Dordrecht, NL: Kluwer Academic Publishers.

van Asselt, M.B.A. and A.C. Petersen (eds.), (2003). *Not Afraid of Uncertainty*, Utrecht, NL: Lemma Publishers. [In Dutch. Published for the Netherlands Advisory Council for Research on Spatial Planning, Nature and the Environment (RMNO).]

van der Sluijs, J.P. (2002). "A way out of the credibility crisis of models used in integrated environmental assessment", *Futures*, **34**: 133–146.

van der Sluijs, J.P., J. Potting, J. Risbey, D. Nuijten, P. Kloprogge, D. van Vuuren, B. de Vries, A. Beusen, P. Heuberger, S.C. Quintana, S. Funtowicz, J. Ravetz, and A. Petersen (2002). *Uncertainty assessment of the IMAGE/TIMER B1 CO_2 emissions scenario, using the NUSAP method.* Bilthoven, NL: Dutch National Research Programme on Global Air Pollution and Climate Change, Report 410200104 (2001).

van der Sluijs, J.P., J.S. Risbey, P. Kloprogge, J.R. Ravetz, S.O. Funtowicz, S. Coral Quintana, A.G. Pereira, B. De Marchi, A.C. Petersen, P.H.M. Janssen, R. Hoppe, and S.W.F. Huijs (2003). *RIVM/MNP Guidance for Uncertainty Assessment and Communication: Detailed Guidance.* Utrecht, NL: Utrecht University. Available at: http://www.nusap.net/downloads/detailedguidance.pdf.

van der Sluijs, J.P., P.H.M. Janssen, A.C. Petersen, P. Kloprogge, J.S. Risbey, W. Tuinstra, and J.R. Ravetz (2004). *RIVM/MNP Guidance for Uncertainty Assessment and Communication: Tool Catalogue for Uncertainty Assessment.* Utrecht, NL: Utrecht University. Available at:
http://www.nusap.net/downloads/toolcatalogue.pdf.

van Egmond, N.D. (1999). "Modellen geven meetresultaten betekenis" ("Models give meaning to measurement results"), *Trouw*, February 3: 16

Walker, W.E., P. Harremoës, J. Rotmans, J.P. van der Sluijs, M.B.A. van Asselt, P. Janssen, and M.P. Krayer von Krauss (2003). "Defining uncertainty: A conceptual basis for uncertainty management in model-based decision support", *Integrated Assessment*, **4**: 5–17.

Weinberg, A.M. (1972). "Science and trans-science", *Minerva*, **10**: 209–222.

CHAPTER 12

TERRY SHINN

WHEN IS SIMULATION A RESEARCH TECHNOLOGY? PRACTICES, MARKETS, AND LINGUA FRANCA

The practices of simulation are highly diverse: In a given domain, and for a single problem, practices are frequently multiple, different from one another, divergent, and more than occasionally contradictory. In view of such acute plurality, how is one to grasp the sense of 'simulation'? While the number and scope of practices embedded in simulation and the magnitude and heterogeneity of the simulation market are synonymous with fragmentation, does fragmentation necessarily prescribe the operation of the simulation community, and if not, what might be the form and function of a said community? Is it reasonable to speak in terms of simulation as a system, and if so, on what grounds? This chapter explores this and related issues. It examines transverse features of simulation that serve as operators of cohesion, which cohesion constitutes a prerequisite for the stabilization of a social/cognitive system. The canvas presented here will include historical, organizational, professional, and epistemological components. The context of the emergence of the C++ general-purpose, multi-paradigm, object-oriented simulation language will be explored. It will be suggested that the intellectual and social dynamics of C++ strongly reflect key features of generic instrumentation and research technologies, and that, by virtue of this correspondence, it is reasonable to think of the practices, structures, and market of simulation in terms of a transverse research technology system.

FOUNDATIONS

Many of the cognitive, organizational, and institutional elements constitutive of contemporary simulation were introduced before 1960. With a few notable exceptions, such as visualization techniques and virtual reality, what has happened since is an extension of that early orientation. Recent change has largely occurred within the confines of the historical mold and logic that initially formed today's huge and diverse domain of simulation thought and action.

Questions of simulation emerged for the first time in Germany in the 1920s when H. Roeder took out a patent for devices intended for use in training pilots of submers-

J. Lenhard, G. Küppers, and T. Shinn (eds.), Simulation: Pragmatic Construction of Reality, 187–203.

ibles, balloons, and airplanes. The simulators were designed to represent changes of altitude in three planes of movement, to register commands initiated by pilot trainees, and to readjust altitude correspondingly. The project came to naught, but the effort is important. It connects simulation to the emergence, development, and currency of new forms of technological artifacts. It is an extension of a historically novel form of technical design, materials, and activities. From the outset, simulation was connected to aviation, issues of complex motion, equations that describe complex flows and interactions, and the extension and adoption of new varieties of skills (sometimes entailing new training programs). Finally, the simulation trainers project of the 1920s established important parameters that underpin thinking and action in most simulation ventures: 1) representations of multidimensional environments, 2) interaction between elements figuring in the representation (sometimes including human operators), 3) emphasis on TIME, that often comprises a key dimension beyond the three axes of freedom in space (particularly in virtual reality), 4) testing, and 5) validation.

However, it was not until the 1940s that Roeder's intuition that the components of simulation, aviation, and training comprise an integrated unit became a reality, and when it occurred, it was not in Germany but instead in the United Kingdom and United States. This gap corresponds to a massive growth of a simulation market in the shape of war-driven demand for expanding numbers of increasingly advanced combat aircraft, for attendant pilots, and for quick efficient training. It similarly corresponds to the design, construction, and spread of simulation-relevant technology such as 'fast' analogue calculating devices, capable of coping with elementary fluid flow equations and their translation into simulation dynamics and mechanical outputs adapted to an aircraft control environment. This evolution proved crucial: Today's faster, better, and generalized simulations rely entirely on digitalized calculations! The historical multifront technological advances of the 1930s and 1940s in electronics and calculation provided the mental and material conditions fundamental to simulation. The centrality of technology to simulation cannot be overestimated. Simulation is the technology of technology, of science, and of industrial operations and beyond. In its role as the technology of technologies, simulation represents the most reflexive form of analysis/action yet known to humanity, and broadly practiced in society.

In 1939, Professor L. Mueller, working at MIT, designed and built a fast analogue computer to study the longitudinal dynamics of aircraft motion. Mueller's interest was in the solution of the set of aerodynamic equations and their simulation for design purposes. However, in a postscript to his paper, he mentioned that his simulator could be adapted to flight simulation pilot training. In 1941, an electronic analogue computer was developed at the TRE unit for the radar training program. The TRE group, originated at MIT during the war, combined advanced detection, electronics, and control systems coupled to fast calculators, mainly for military objectives. This device was based on the ideas of F. Williams, famous for his later contributions to digital computers, and used the velodyne – another TRE invention for integration. The first model of this computer had been developed by Dynatron Radio Ltd. in 1941.

The war years saw an increase in the companies involved in simulation as well as in hardware, accessory, and modeling technology. In 1945, a new system was introduced by K.M. Uttle that incorporated force fields and visual inputs into flight trainers. In Britain, advanced longitudinal dynamics was added to the flight simulation repertory through the efforts of G.M. Hellings and the resulting electromechanical analogue computer. This computer and model was arguably sufficiently general and flexible to correspond to the flight characteristics of any aircraft then in operation. The simulator boasted a pitch motion system that incorporated an endless moving belt. In the United States, the Special Devices Division of the Bureau of Aeronautics developed the Center for Naval Training Equipment. It supported the simulator activities of Bell Laboratories, which constructed the Navy's PBM simulator that included a complete front fuselage, cockpit, accessories, and all instrumentation. The Link trainer was also developed during this period. In 1948, and during part of the 1950s, the Curtiss-Wright aviation firm engaged in simulator design, The company produced a new line of servo devices and what was known as shadow graphics. General Electric entered the simulation race during the 1960s, developing digital systems for space-related operations.

Work surrounding the atomic bomb constitutes a second current of simulation activity. During the latter phase of bomb research, Los Alamos scientists and engineers set out to study the magnitude of their bomb's explosive impact. At the time, nothing was known about nuclear blast extent. To determine this, scientists selected numerous possibly relevant parameters, and assigned a huge variety of values to each. The number of permutations was astronomic. They engaged the newly emergent computational technology of computers becoming available at that timc to calculate the likelihood of the selected parameters and to estimate the consequences of each. The resulting calculations indicated a statistical likelihood of bomb effects. The probability-based technique of this project soon acquired the name *Monte Carlo simulation*, presumably suggesting the probabilistic aspects of operations. It rapidly became central to a sweep of simulation endeavors. On a different register, starting in the 1950s with scenarios based on 'if-then' logic, insurance companies used Monte Carlo simulation to work out actuarials, and soon banks and investment firms were using it for investment and client advice. Monte Carlo simulation has similarly become the cornerstone for much risk assessment research and public policy, and it is today a technique deployed by nuclear energy lobbies seeking to quiet public unease about nuclear hazard. The introduction of high-power individual computer technology has even further accelerated the generalization of this form of simulation (e.g., for calculating possible trends in the stock market).

The initial organization of simulation technology was rooted in engineering practice and undertaken by engineers and not by scientists, scientific societies, or universities. One of the first simulation forums to be scheduled and the first simulation organization founded was venued in Europe, and not in the United States, where much of the early simulation work had been carried out. One explanation for this is that part of European postwar reindustrialization was free to build around new technology, the older prewar production capacity and technology having been destroyed by bombing, battle, and sabotage. In 1955, a meeting was convened in Brussels at the Free University attended by researchers, managers, and observers of the simulation laboratories that existed at the time. Many of the laboratories were a by-product of the enormous simulation-related activities of World War II. Participation was international, with delegates coming from most west European countries, the United States, and Japan. The conference decided there existed a need for a permanent means of communication between members of the emerging simulation community. Participants perceived that extant professional and scientific bodies were too restrictive in composition and outlook to include the diversity of backgrounds, skills, and interests that characterized the multidisciplinary, multisectoral, and multipractice new world of simulation. The result was the creation of the AICA (*L'Association Internationale pour la Computation Analogue*). Under the influence of technological and scientific innovation, the initial scope of the AICA expanded to include more mathematical analysis (particularly numerical operations), mathematical modeling, and digital technology. This body played a crucial role in the introduction of simulation particularly into Europe. Its outstanding successes lay in the domains of chemical engineering, automatic system engineering, and later simulation-based design, specifically in the realm of mechanics. In 1976, the pioneering body took the name *International Association for Mathematics and Computation for Simulation*, to better reflect the broadening technology, uses, and markets of simulation. The original AICA was significant for the establishment of simulation, because it tried to coordinate and combine the analytic practices developing in simulation, which, at the time, was still an outside cognitive corpus and set of practices as well as being socially nebulous.

SIMULATION AT WORK – POST 1960

Most post-1960s simulation activity has focused on engineering-related and industry/service-oriented work, as measured by the focus of simulation societies and journals. John McLeod and Vincent Amico have been pillars of simulation in the United States since World War II. McLeod earned a BS in engineering at Tulane University, and has been associated with the universities of Chicago, Harvard, and MIT. He is an expert in the design and construction of automatic control systems (boasting two patents) in which simulation is the principal tool. He served in the US Navy's Guidance Systems Simulation Laboratory for a decade, acquiring initial simulation-based design experience there. He then went to work for the Northrop Company, moving to General Dynamics between 1956 and 1963 – again in design. McLeod became an independent simulation researcher for a brief period during the 1960s when he designed a heart-lung machine using principles of simulation-driven design techniques. He was an active consultant throughout his career.

Amico earned a BS in engineering from New York University in 1941 and went on to study physics. From 1941 to 1945, he worked on the structural design of missiles and aircraft at the Static Test Laboratory at Wright Field. After 1948, he worked as a civilian for the US Navy as product engineer for flight training equipment. In 1969, Amico became Research Director of engineering of his design unit; and, in 1979, he was appointed Research Director of the Navy's entire flight training program. Throughout, Amico based his endeavors on simulation development and simulators. They comprised the mainstay of his career. Beginning in 1972, he taught at the computer department of the University of Southern California, specializing in simulation techniques.

McLeod and Amico figured importantly in the development and successes of the Society for Computer Simulation that became the United States major organization specializing in simulation publications, and to a lesser degree simulation conferences. Although the Society was set up in 1952, it remained a relatively small and obscure body until the late 1960s, when McLeod helped expand its activities, often assisted by Amico. McLeod turned this American organization into an international body; and under his careful guidance, the Society was rechristened the *Society for Modeling and Simulation International.* McLeod managed to connect the American simulation community to a broader simulation environment based in Great Britain, France, Holland, Germany, Italy, Denmark China, Korea, and Japan. From the 1970s onward, simulator experts and users regularly participated in the annual meetings of the Society for Modeling and Simulation International – the *Summer Computer Simulation Conference* that specializes in continuous event simulation.

The Society for Modeling and Simulation operates a large, prestigious stable of simulation publications. It publishes *Simulation* – the society's flagship review, and the most well-read and respected journal in the field. It also publishes *Transactions for Modeling and Simulation*, *Simulation Magazine*, and *The Journal of Defence Modeling and Simulation: Applications, Methodology, and Technology* (with the US Army and Simulation Office).

However, it is not the Summer Computer Simulation Conference that constitutes the foremost venue for the exposition of new work in the field of simulation, but instead the *Winter Simulation Conference.* Whereas the beginnings of the Winter Simulation Conference date back indirectly to a series of small simulation seminars held during the late 1940s, the organization was set up only in 1967. The initial meeting was headed by H.J. Hixson (head of operations systems research analysis with the US Air Force logistics command and program director of the IBM SHARE users group) and by J. Reipman (prominent user of the general purpose simulation system approach in the Nordon division of the United Aircraft Corporation and a leader in the IEEE). It differs from the Summer Computer Simulation Conference in several important respects. The latter deals with continuous simulation, whereas the Winter Conference specializes in discrete event simulation. Discrete event simulation efficiently represents events in which time is a subordinate consideration. In contrast, continuous simulation, based on the solution of differential equations, unceasingly monitors time, but is less attentive to details of complex events. Continuous simulation is used to model activities like continuous flow engineering and aircraft automatic pilot systems. Manufacturing and services are modeled with discrete event simulations. However, as indicated by Küppers and Lenhard (this volume), a closer analysis reveals that matters are becoming increasingly complicated, because even continuous simulation relies on discrete models.

Whereas the Summer Conference has rejected sponsorship by professional engineering bodies, interest groups, and public agencies, the Winter Simulation Conference has multiplied such connections. The Association for Computing Machines, the IEEE, and IBM sponsored the initial meeting. An audience of 225 was expected, but interest ran so high that 401 attended. The proceedings were published by the *IEEE Transaction* in a special issue in 1968 on systems science and cybernetics.

The second Winter Simulation Conference took place in December 1968: Its theme was *Simulation Applications*. In addition to the initial sponsors, this meeting also received backing from Simulations Council Inc. There were twenty-two sessions, and eighty-eight papers were presented on a range of simulation applications. Statistical research, simulation computer language development, and simulation education were included in the program. Attendance jumped to 856. A 356-page digest of conference papers was subsequently published. The 1969 Winter Simulation Conference was sponsored by the American Institute of Industrial Engineers and the Institute of Management Sciences/College on Simulation and Gaming. In 1971, the Operations Research Society of America also became a sponsor. That year, attendance reached over 1,200 – the highest figure ever.

In 1974, the tide turned: The number of participants fell sharply, as did financial support for the program. The future looked bleak for several years, until the National Bureau of Standards intervened, infusing organizational vigor, new ideas and projects, and fresh money. In the 1980s, the Winter Simulation Conference regained its former ascendancy and has maintained it ever since. But, why the collapse in the mid-1970s, and what structural considerations contributed to its newfound energy?

Although the exact circumstances require further research, one can point to several contributory factors: By the 1970s, there existed a plethora of simulation research directions, projects, application niches, and implementations. The field had fragmented considerably. Functional sectors and individual firms were working out their specific simulation solutions. The initial flush of enthusiasm that fuels a new venture had begun to erode. Two key initiatives regalvanized the simulation venture. Intervention by the US National Bureau of Standards provided a measure of stability and coordination that was otherwise lacking. The Bureau pulled together divergent simulation movements. Second, new initiatives in programming language began to emerge. Introduction of a language like C++ (to be analyzed in detail below) helped the simulation community by providing a focus of technological, intellectual, and professional convergence.

In sum and as indicated above, military-related programs lay at the center of the development of simulation work. This was facilitated by the introduction of fast digital computing power and by the swift broad spread of computers. Nevertheless, simulation efforts had begun to thrive on the basis of predigital computation. Slow analogue devices had already permitted simulation to successfully invade a growing range of military-related realms before advanced digital developments. Additionally, the vast majority of simulation endeavors occurred in the narrow engineering/technology sphere.

STRUCTURING SIMULATION – THE BIRTH, EVOLUTION, AND ROLE OF THE C++ GENERAL-PURPOSE, MULTI PARADIGM, OBJECT-ORIENTED PROGRAMMING LANGUAGE

The pages that follow will document the centrality of the C++ computer language to the internal development and point to the diversification and growth of the simulation markets and community since the mid-1980s. It will be further suggested that C++ exhibits many of the key attributes of research technologies. Grounded on these twin observations, I argue that simulation itself may usefully be perceived as a research technology rich in generic instrumentation that simultaneously provides a stable kernel to the simulation domain and permits diversity, yet transversality, commensuration, coherence, and cohesion.

The C++ simulation-directed, general-purpose, object-oriented, multipurpose language was developed between 1983 and 1985. The programming language's author is Bjarne Stroustrup (1950), a brilliant, energetic, successful, and some would say charismatic, general-purpose language developer. Today, C++ is the most widespread language in simulation. In 2003, estimates range between 1.3 and 3 million users. C++ is a generic technology, fathered, matured, and organized in an interstitial environment. The disembedding of its generic features and their reembedding in specific applications involves intermittent selective boundary crossings. It corresponds to a form of metrology. By virtue of this combination of characteristics, C++ constitutes a research technology (Joerges and Shinn 2001; Shinn and Joerges 2002; Shinn and Ragouet 2005).

Bjarne Stroustrup is currently the College of Engineering Professor at the Texas A&M University Department of Computer Science, and director of the Large-Scale Programming Research Department at AT&T Laboratories. He was born in Denmark, and did his undergraduate work at Aarhus University, taking a degree in mathematics and computer science, before moving to Cambridge University for his doctoral studies. His dissertation adviser was David Wheeler, a well-known programmer who contributed to the Illiac. The Illiac, built in 1952, based on von Neumann architecture and located at the University of Illinois, was the United States' most powerful university computer, even surpassing the capacity of the combined Bell Laboratory machines.

Stroustrup studied at the Cambridge Computer Laboratory, conducting research on alternatives for the system software of distributed systems. He composed new software from existing systems and tested feasibility and efficiency using simulation techniques. On completing his doctorate in 1979, Stroustrup took a position in the computer science research center in Bell Laboratories at Murray Hill, New Jersey, where he undertook research alongside language specialists like Denis Ritchi, who had recently developed the programming language C. From 1979 to 1983, when Stroustrup set out to build a new language, he was involved in a range of Bell-related tasks. He was notably active in simulation research intended to improve distributed network system operations, and explored applications of this approach. In 1979, Stroustrup set out to analyze the Unix kernel to determine how it could be distributed over a network of computers connected by a local area network. He also worked on improving the low-level language C (Stroustrup 1993).

One way of describing C++ is that it contains many elements of C that have been enriched with Simula and an object-oriented perspective. Stroustrup often says that C++ is three languages in one: a C-like language (supporting low-level programming), an Ada-like language (supporting abstract data-type techniques), and a Simula-like language (supporting object-oriented programming) (Stroustrup 1994: 198). C++ is also organically connected to additional languages (Stroustrup 1994: 198). Algora68 gives to C++ operator overloading and the capacity to declare variables anywhere in a block. BCPL allows comments. Simula gives organization.

Whereas one strength of C is its proximity to computational machines, via the introduction of elements from Simula and from the object-oriented perspective, C++ connects directly to material problems by analysis of language application. C's logic connects with computational machinery, whereas C++'s logic retains this property and adds the property of smooth problem application logic.

Stroustrup's design emphasized three stable features – application, the generic concept of classes, and portability (the latter allowing cross-boundary flexibility and translanguage communication). Stroustrup has always been concerned with promoting solutions to real problems (Stroustrup 1997a,b). While his Denmark training in mathematics was interesting and stimulating, it nevertheless left him uncomfortable, as he is committed to confronting problems. He often insists that his language insights and successes result from thinking about programming with reference to personal problem-solving experience. He generalizes up from problems. C++ has been built to enable real and diverse users to better grasp their problems and treat them computationally. In this sense, the C++ language is application- and user-driven.

Second, C++ was designed to operate in a framework of classes. Initially, C with classes was developed by Stroustrup to allow simulators to be built for research in network design being carried out by Sandy Fraser at Bell. Inclusion of the term 'C' in the name C++ indicates the extent of C's parentage to C++. One frequently asked question is why C++ did not simply emerge as an evolution of C, rather than distinct from it, and as a powerful and eventually victorious competitor. Part of the answer is the centrality of classes in C++. While Stroustrup seriously tried to reconcile the importation of classes into C, as witnessed by his construction of C with classes, the architecture of C limits the full expression of classes. This drove Stroustrup to further diverge from C, as he continued in his project to build a more useful programming language. The relationship between C and C++ is expressed in a phrase submitted by Stroustrup in a 1989 article *As close to C as possible, but no closer* (Koenig and Stroustrup 1989).

According to Stroustrup, classes possess a multifold advantage (Venners 2003; Dolya 2003). Classes promote reasoning in terms of connections crucial to object-oriented representations and work. C++ is not an object-based programming language; it is an object-oriented code, which may be more restrictive. Classes help identify the similarities shared by elements. They facilitate computation between them. They furthermore allow passage from one part of a program to another with a minimum of difficulty, facilitating the work of programmers and users. Classes in a program reduce runtime, making computation efficient. In C++, the combination of classes and static checking helps alleviate the need for garbage collection, which constitutes an important economy in runtime and memory. Stroustrup also suggests that elegance is a desirable quality of a good language, and the use of classes promotes elegance. But, above all, in C++, the true purpose of classes is that they provide a platform for clear reasoning about complex structures.

Classes, in conjunction with other components like static checking, restrictive garbage collection, and multiple inheritance, constitute the generic feature of C++. The centrality of classes in C++, along with the role played by the object-oriented perspective, makes C++ ubiquitous in much simulation work.

While classes comprise the generic anchor of C++ that allows the general purpose code to be adapted and adopted by multiple applications, portability represents the format through which classes and other generic features of the language are vehicled. Portability was crucial to Stroustrup from the outset. He tried to design a language far more portable than C. Portability infers compatibility and flexibility. C++ is not a restrictive language, as it can easily be connected to a range of hardware and software. It operates in innumerable environments – Unix, Windows, and Macintosh. C++ links conveniently to many application tools. It is thus embedded in a huge range of informatic products without restriction, whereas the architecture of other languages limits their partnering. C++ functions as a foundational language tool inside an ever extending variety of application-specific local tools. Most computer languages are written either by language designers for other designers or by users for a particular application; but this is not the case for C++, whose architecture integrates a breadth of perspectives and conveys a multitude of mechanisms felicitous to mobility. This drive for breadth echoes Stroustrup's twin concerns for and experiences in concrete application practice and design involvement.

Portability is the hallmark of boundary crossing. This feature permits the expression of classes in terms of genericity and class reembedding in diverse applications. Without C++'s portability, movement across boundaries would be rare or impossible. Portability spells ongoing communication between evolving C++ design and otherwise isolated C++ niche users. Thanks to portability, C++ can thus stand as a generic language, as a language of application, and also as a reflexive transverse language that permits transapplication exchange.

The break of C++ with C, and its subsequent promotion in 1984–85, proved both problematic and easy. Its possible competitors, Modula-2 and Ada among others, were often regarded as restrictive, entailing awkward problems, or simply did not find dynamic outlets on the United States market. The Bell Laboratory, where Stroustrup worked, had experimented with and contributed to numerous languages, and was thus not irreconcilably committed to any particular one. The research unit was big, leaving room for individual initiatives and maneuver. Furthermore, Dennis Ritchi, who also worked at Bell and, along with Kristen Nigaart, is one of the pioneers of C, never strongly opposed Stroustrup and his endeavors, which increasingly distanced C++ from C. The rapid successes of C++ also owes much to the expansion of the mini- and microcomputer market and to the growth in the number and range of applications. Foremost in these applications was simulation, to which C++ was perceived as appropriate and congenial. The diffusion of C++ was also connected to keen interest in the new language among language designers and programmers. C++ required good compilers and libraries to ensure its spread and effectiveness in different applications (Venners 2003), and much to even Stroustrup's surprise, the computation community responded in record time to his proposed architecture with a number of world-class compilers.

The number of C++ users rocketed. In 1984–85, the emergent language was largely restricted to the Bell Laboratory. In the months that followed, C++ was distributed in a preliminary version to selected universities and a few users (Stanford, University of California, Cal Tech, University of Wisconsin, MIT, Carnegie Mellon, University of Copenhagen, Rutherford Laboratory in Oxford, etc.). The response was not what was expected. Rather than expanding, demand stagnated. The motive for this was unanticipated. Users loved the new language and wanted nothing better than to use it extensively: but its use in consulting and in public applications required the stabilization and standardization of the language, its public recognition, and the development of adequate compiler architecture. Soon, however, the quick involvement of fresh C++ work, often originating outside of Bell Laboratories, allowed both the size and diversity of the C++ community to expand on an unforeseen scale.

Table 1. Growth in the size of the C++ community

Date	Estimated number of C++ users
1979	1
1980	16
1981	38
1982	85
1983	87
1984	135
1985	500
1986	2,000
1987	4,000
1988	15,000
1989	50,000
1990	150,000
1991	400,000
2002	1.300,000

The following ventures indicate the range of activities associated with C++ in the decade since 1985: Animation, autonomous submersibles, billing systems, bowling alley control, circuit routing (telecom), CAD/CAM, chemical engineering process simulations, compilers, control panel software, cyclotron simulation and data processing, database systems, decision support systems, digital photography processing, digital signal processing, electronic mail, expert systems, factory automation, financial reporting, flight mission telemetry, foreign exchange dealing (banking), search software, hardware description, hospital records management, industrial robot control, instruction set simulation, interactive multimedia, magneto hydrodynamics, medical imaging, missile guidance, mortgage company management, network management and maintenance systems (telecom), network monitoring (telecom), operating systems (real-time, distributed, workstation, mainframe, 'fully object-oriented'), programming environments, superannuation, insurance, shock-wave physics simulation, SLR camera software, switching software, test tools, transmissions systems (telecom), transport system fleet management, user interfaces, video games, and virtual reality (Stroustrup 1994: 172).

By the late 1980s and 1990s, C++ had become a major language among computer languages, or perhaps even the foremost. The *C++ Journal* appeared in 1991. *Computer Language*, *The Journal of Object-Oriented Programming*, *The C++ Users Journal*, *Journal of Object-Oriented Programming* (*JOOPS*), and *Dr. Dobbs Journal* all ran regular articles on the C++ language. To this must be added the score of Internet language publications that frequently feature C++ and present recent C++-related tools and implementations.

C++ owes much to social factors. By the mid-1980s, the computer language and computer programming community had grown greatly. The consequence was twofold: First, numbers of young talented specialists could now loosen their ties with the big powerful computer firms that had earlier exercised considerable influence. Second, conditions permitted some individuals to become independent free-lance programmers. In 1986–87, a movement of independent compilers developed. This was opportune for C++, whose evolution and diffusion required expanding beyond the walls of Bell laboratory, and the participation of many people and inputs from many quarters.

The now famous Santa Fe compilers meeting was held in November 1987. This event marks a turning point for C++. Stroustrup anticipated an attendance of only a few dozen people in Santa Fe, but over 200 showed up! Papers were presented on C++ application, education, environment compatibility, and, strategically most crucial, on compiler development. Building effective C++ compilers was essential to users, as they ease the work of application. For a compiler to be in step with the C++ generic/reembedding/boundary-crossing Stroustrup precepts, it must support many other operating languages. This meeting initiated the design, construction, and diffusion of a spate of C++ inspired compilers and libraries – among others, the 1988 Zortec compiler and the 1990 Borlan compiler. In 1991, Windows marketed its C++ compiler; and in 1992, IBM came out with its version of a C++ compiler. A C++ groundswell ensued. In 1988, the NIH helped sponsor a C++ meeting and began to acquire C++ programs, tools, and other application implementations. In 1988, the second C++ conference was held In Denver, and, since then, there have been C++ conferences on an almost yearly basis.

During the 1980s, why did Bjarne Stroustrup push first to obtain the standardization of C++ by ANSI (the American National Standards Institute) and then by ISO a decade later (ISO/IEC14882)? What does standardization signify for a code, and what form of work is involved? How does standardization impact on a language? What has the standardization of C++ meant to simulation?

Before the standardization of C in the late 1980s, the language counted over 160 dialects. In this instance, standardization neutralized fragmentation and imposed order. The development of libraries and compilers can be used to hijack a language by locking in users. By standardization, a code becomes public and thus cannot be appropriated. Stroustrup deplores the idea of a proprietary language, and above all desired C++ to remain public. His design goal and subsequent strategy entailed that

C++ be an open pathway, not a closed system. He has often declared that C++ is for the average user as well as for the untypical user. It is for everyone! Finally, the documentation that necessarily accompanies standardization allows clarifications – clarifications in the work of design itself, design improvements, and clarification for users who want or need to know more about their code. Metrology thus fulfills the functions of stabilization, transparence, accessibility, and pedagogy.

In the early 1990s, C++ was certified by the *American National Standards Institute*, the *German Institute of Norms*, and by the *British Institute*. This achievement culminated five years of effort. Stroustrup was motivated by numerous considerations during his drive to win certification. Although the growing number of users approved and employed his language, general acceptance demands code homogenization and stabilization. This is necessary for a technology to be perceived as transparent, transferable, and reliable. Stroustrup understood this. By all means, code deviance must be avoided and prohibited. Lock in of parts of the language also had to be prevented! Standardization of C++ at ANSI required several years. ANSI operates in conjunction with code designers, subsequent contributors, tool and implementation designers, firms that develop and market the tools, attorneys of said firms, independent code experts, and users from many sectors. Observers and freelancers are also implicated in the standardization process. Deliberations occur on two levels: Technical committees deal with issues of detail. A general body discusses questions of principle, law, and policy. For C++, deliberations advanced relatively smoothly. At the end of the process, Stroustrup had given up nothing essential to his initial design plan.

The ANSI and ISO procedures affected several important evolutions in C++. Templates were made central to the code. They were connected to libraries, and the template standard (TSL) resulted. Multiple inheritance also became a key feature. Stroustrup had earlier been weary of inheritance, but in the form that it was engineered during standardization and combined with C++, he came to accept it and appreciate its power. The next steps in modifying C++, states Stroustrup, will be its extension to distributive programming that will necessarily introduce threads (Dolya 2003).

The upshot of standardization is that C++ became even more general-purpose/multiparadigm than before. C++'s compatibility with other codes was enhanced. It is a general user's language, a specialty language, and a programmer's code. It remains a language for high-level designers, as its openness and generic quality continue to make it interesting, challenging, and a turf still sufficiently malleable and open to correctly support future evolutions.

THE INTERSTITIAL ENVIRONMENT

Genericity, reembedding, and boundary crossing are coupled to an interstitial environment, and this environment figures centrally in the development of C++ practices and the simulation community. An interstitial arena emerges in the interspecies between established dominant organizations, such as the university, corporations or small technology-based firms, state technical services, the military, and so forth. While individuals who occupy the interstitial environment may work for a dominant organization, they nevertheless frequently escape organizational control by movement or by fostering arrangements that connect them to multiple organizations. Multiple connections increase resources and extend margins of maneuver. He who works for everyone is the bondsman of no one. In what ways has Bjarne Stroustrup and the design and evolution of C++ been affiliated with the interstitial environment?

Today, Stroustrup divides his time between contacts with C++ users and markets, code design, education, and acquiring new language skills and experience (Doyla 2003).

Stroustrup's experience is distributed, extending to pure and applied mathematics, language and program writing and use, and simulation. His endeavors bridge theory and application. Some of this output occurs in the framework of universities, state research or technical services (the NIH), private/semi-public services (ISO and ANTI), huge corporations (Bell and AT&T), and small technological companies (Silicongraphics). Concurrent and sequential employment is a rule. While once an employee of Bell Laboratories and still an active employee of AT&T Laboratory, Stroustrup has often simultaneously held other positions.

The interstitial environment respects and maintains divisions of labor inside science, inside technology, between science and technology, and between science and enterprise, while new important permutations are also invented continually (Joerges and Shinn 2001; Shinn and Joerges 2002; Shinn and Ragouet 2005). Stroustrup held to his C++ general-purpose, multiparadigm project. He used the relative autonomy provided by the interstitial environment to focus long-term attention on his objective. The interstitial arena is not an interest group, not a producer, and not a market. As often stressed by Stroustrup, his goal is steadfastly nonproprietary. He has no plan to encroach on the productions or operations of specific user/market niches. Consistent with this neutrality, the interstitial arena thus provides Stroustrup a platform for crossing the boundaries of innumerable organizations, markets, producers, and users, but without affecting the internal division of labor or infringing on their traditions, plans, or autonomy. He reembeds C++'s generic features into niche applications, and conversely uses the application experience of niches to design and enrich C++.

C++ comprises a lingua franca: A lingua franca is the result of genericity, reembedding, and boundary crossing. A generic technology like C++ contains one or several fundamental instrumentation features, for example, an emphasis on classes in combination with abstraction, object orientation, and portability. When the generic technology moves into a particular market niche, several things occur. The generic features of the technology are reembedded in the local technical culture. In the case at hand, applications absorb certain selected features of C++ in accordance with short-term demand. Parts of C++ are reshaped in this process of adoption. Nevertheless, C++'s adaptation does not alter the fact that the generic characteristics of the base technology survive intact. The stamp of the generic base is permanently imprinted on the local technical expression. The lingua franca arises out of this complex concatenation. For example, a specific library accompanies C++ into the separate applications of hydrodynamics research, auto piloting, and so forth. Nevertheless, the presence in C++ of a base code governing the organization of classes and logic of portability persists both in the piloting and science research extensions, and it is these stable, constant transverse features that enable users from both specialties to communicate effectively about the language and about more substantive issues beyond their fields on a metalevel.

The reembedded generic technology vehicles a particular metrology in the form of standards or units of measurement, a specific vocabulary, form of imagery, methodology, or even a new paradigm. This residue is deposited in the local niche technology during reembedding. Although the product of reembeddings in diverse markets results in heterogeneous artifacts, the generic element remains uniform, a kind of technical fingerprint. However, C++ is expressed in the innumerable applications it serves; C++'s underlying signature persists. Through 'assimilating' the metrology, methods, vocabulary, or images of the hub generic instrument, niche practitioners come to share familiarity with, and competence in, a particular syntax and semantics. This common language becomes an integral feature of the practitioner's niche language. The language is associated with a set of local, efficient, robust practices. The technology, practices, and outcomes associated with the local techniques enjoy the status of 'truth' – in the sense of 'practical truth.' The lingua franca of C++ is the generic residue of the reembeddings of C++ in a multiplicity of fragmented market niches. It is that part of the C++ hub technology that transcends the transformations occurring during the process of adaptation and adoption. Since there persists a transverse stable kernel used in the discourse of market practices, when practitioners who come from diverse economic or disciplinary sectors and have different functions (and come from different nations and have even different cultural horizons) meet, they can nevertheless communicate with reference to a technical field and generate intelligibility. The common parlance available through this generic-driven lingua franca promotes cognitive, artifactual, and organizational transversality that somewhat neutralizes the otherwise often disruptive effects of today's rampant intellectual and social differentiation and fragmentation. Witness to the existence and efficiency of C++ as a lingua franca in simulation can be found in two venues. The annual Summer and Winter Simulation Conferences draw users from scores of applications. They communicate through C++ about C++, and also use the medium of C++ to communicate about their respective different and sometimes divergent simulation applications.

By virtue of the fact that C++ represents a kind of research technology connected to many and diverse audiences and functions and highly amenable to transversality in the form of boundary crossing and commensuration, and by virtue of the fact that C++ constitutes a dominant code in simulation practice, one may reasonably establish the operation of a link between simulation practices and markets on the one hand, and research technology on the other. Simulation's stability and strengths owes much to atributes drawn from research technology. Research technologies provide simulation (here demonstrated through analysis of C++) with open-ended techniques, representations, codes, and language that make it applicable in a miriad spheres, and it simultaneously offers a solid, self-referencing platform that gives definition, meaning, and direction to the more general, overall, quasi-universal simulation enterprise.

WHEN IS SIMULATION A RESEARCH TECHNOLOGY?

Simulation is a system whose architecture emphasizes three features: First, simulation involves a remarkably large number of markets, and the number continues to expand. Second, the particularities and demands of simulation niches are diverse and even divergent. Many markets require specific simulation practices, representations, robust results, and efficiency. Engineering demands predictability and information that promotes risk avoidance. Science counts on precision and understanding. These demands on simulation are very different. Third, traversing all markets and practices, there exists a kind of common denominator of expectations. Practitioners demand 'truth.' The relevant form of truth might be described as 'practical truth.' Practical truth, as distinguished from the epistemological truth of philosophy, refers to dependable matter-ground individual and group satisfaction based on perceived reliable material inputs and outcomes. What counts here as 'satisfactory' is connected both to individual experience and to the expectations and norms of society. Practical truth is hence simultaneously concretely personal and the fruit of collective routines. Upon determining that technical objects and protocols yield the anticipated result, practitioners gradually come to equate a technical ensemble as 'true' to their understanding of the technical system's properties and ensuing yield. Indeed, stable controlled technical 'yield' comprises the key facet of pragmatic truth.

Heterogeneity of environment, form, and function are constitutive of the underlying logic and action of simulation. In view of the extremes of heterogeneity, what gives the simulation system substance, stability, and continuity? Part of the answer lies in research technology.

Several of the principal ingredients of research technology have been introduced in the above analysis: genericity, the interstitial environment, markets and technical niches, divisions of labor, relative autonomy, boundary crossing, reembedding, commensuration, practical truth, and lingua franca. Simulation is stretched between two imperatives: its multiple practices and markets; and the need to preserve a stable and standard kernel. Simulation necessitates unchanging standards in order to convince users that outputs are effective and comparable, and to provide ground rules for internal community communication and the further development of simulation techniques.

An intellectual and organizational formula must be evolved capable of ensuring transversality across practices and markets while preserving technical and community cohesion. Research technology contributes to this end. In the case of simulation, the emergence of a generic, general-purpose, multiparadigm code specifically designed with objects in mind, like C++, provides a balance between centrifugal and centripetal action. Spawned along tenets of research technology, C++ has given simulation the power to expand into ever more applications and markets through implementation tools (classes, portability, compilers, and standard template libraries). The ANSI and ISO standardization of the code has ensured uniformity and continuity, urgently called for by clients. The standardization process has allowed simulation practitioners and design specialists to further work out generic foundations. Finally, simulation boasts an interstitial environment, witnessed in the complex cognitive and organizational trajectories of Bjarne Stroustrup, John McLeod, and Vincent Amico.

GEMAS, Maison des Sciences de l'Homme, Paris, France

REFERENCES

Allison, C. (1996). Interview with Bjarne Stroustrup on his reaction to the imminent completion of the ANSI/ISO C++ Standard, *The C/C++ Journal,* October 1996, **14** (10); http://public.research.att.com/~bs/interviews.html (acc. April 13, 2004).

Dolya, A.V. (2003). Interview with Bjarne Stroustrup for the *Linux Journal*, posted August 28, 2003, http://public.research.att.com/~bs/interviews.html (acc. April 13, 2004).

IEEE Computer (1998). "Open Channel" Interview with Bjarne Stroustrup: The Real Stroustrup Interview. "The Father of C++ explains why standard C++ isn't just an object-oriented language", http://www.research.att.com/~bs/ieee_interview.html (acc. April 13, 2004).

IEEE Transaction (1968). Special Issue on Systems Science and Cybernetics, Vol. SSC-4, No. 4.

Joerges, B. and T. Shinn (eds.) (2001). *Instrumentation Between Science, State and Industry*, Dordrecht, NL: Kluwer Academic Publishers.

Kalev, D. (2001). "An interview with Bjarne Stroustrup", *LinuxWorld.com*, August 2nd, 2001, http://www.itworld.com/AppDev/710/lw-02-stroustrup (acc. April 13, 2004).

Koenig, A. and B. Stroustrup (1989). "C++: As close to C as possible – But no closer", *The C++ Report*, July 1989.

McLeod, J. (1968). Simulation: The Dynamic Modeling of Ideas and Systems with Computers, New-York: McGraw -Hill.

McLeod, J. (1982a). *Computer Modeling and Simulation: Principles of Good Practice*, The Society for Computer Simulation International.

McLeod, J. (1982b). "Simulation" in the Encyclopedia of Science and Technology, New-York: McGraw-Hill.

McLeod, J. and P. House (1977). *Large-Scale Models for Policy Evaluation*, New York: Wiley-Interscience.

Nelson, E. (2004). Interview with Bjarne Stroustrup for *Visual C++ Developers Journal (VCDJ)*, http://www.research.att.com/~bs/devXinterview.html (acc. April 13, 2004).

Pradeepa Siva C. (2004). Interview with Bjarne Stroustrup *for Addison-Wesly Longman* about the third edition of The C++ Programming Language, http://www.research.att.com/~nineteen nineteen bs/3rd_inter1.html (acc. April 13, 2004).

Shinn, T. and B. Joerges (2002). "The tranverse science and technology culture: Dynamics and roles of research technology", *Social Science Information,* **41** (2): 207–251.

ShinnT. and B. Ragouet (2005). *Controverses sur la science: Pour une sociologie transversaliste de l'activité scientifique*, Paris, Raisons d'agir.

Stroustrup, B. (1980). "Classes: An Abstract Data Type Facility for the C Language", *Bell Laboratories Computer Science Technical Report* CSTR-84.

Stroustrup, B. (1981). "Extensions of the C language type concept", *Bell Labs Internal Memorandum.*

Stroustrup, B. (1982). "Adding classes to C: An exercise in language evolution", *Bell Laboratories Computer Science Internal Document.*

Stroustrup, B. (1986) *The C++ Programming Language*, Reading, MA: Addison-Wesley.

Stroustrup, B. (1987a). "The evolution of C++: 1985–1987", *Proceedings USENIX C++ Conference*, Santa Fe, NM.

Stroustrup, B. (1987b). "Possible directions for C++", *Proceedings USENIX C++ Conference*, Santa Fe, NM.

Stroustrup, B. (1989). "Standardizing C++", *The C++ Report,* Vol. 1, No. 1.

Stroustrup, B. (1990). "On language war", *Hotline on Object-Oriented technology*, Vol. 1, No. 3.

Stroustrup, B. (1993). "The history of C++ : 1979–1991", *Proceedings ACM History of Programming Languages Conference (HOPL-2). ACM SIGPLAN Notices.*

Stroustrup, B. (1994). *The Design and Evolution of C++*, Indianapolis, IN: Addison-Wesley.

Stroustrup, B. (1997a). "The object magazine online interview", http://www.research.att.com/bs/omo_interview.html (acc. April 13, 2004).

Stroustrup, B. (1997b). *Design and Use of C++ by Bjarne Stroustrup*, Talk given November 3, 1997 at Computer Literacy Bookstore in San Jose, CA, http://www.research.att.com/~bs/Cbooks_QnA.html (acc. April 13, 2004).

Stroustrup, B. (2000). *C++ Answers From Bjarne Stroustrup*, e-mail interview with slashdot.org in the week of February 25, 2000, http://www.research.att.com/~bs/slashdot_interview.html (acc. April 13, 2004).

Stroustrup, B. (2004). Interview, http://members.shaw.ca/qjackson/writing editing/interviews/Bjarne-Stroustrup.htm (acc. April 13, 2004).

Venner, B. (2003). "The C++ style sweet spot. A conversation with Bjarne Stroustrup, Part I", October 13, 2003, http://www.artima.com/intv/elegance.html (acc. April 13, 2004).

"Modern C++ style. A conversation with Bjarne Stroustrup, Part II", November 24, 2003, http://www.artima.com/intv/modern2.html (acc. April 14, 2004).

"Abstraction and efficiency. A conversation with Bjarne Stroustrup", Part III", February 16, 2004, http://www.artima.com/intv/abstreffi2.htm (acc. April 14, 2004).

"Elegance and other design ideals. A conversation with Bjarne Stroustrup, Part IV", February 23, 2004, http://www.artima.com/intv/elegance.html (acc. April 13, 2004).

List of Authors

Marcel Boumans, Dr. ir.
Department of Economics
University of Amsterdam
Roetersstraat 11
NL-1018 WB AMSTERDAM
The Netherlands

Michael Hauhs, Prof. Dr.
Ecological Modeling
University of Bayreuth
D-95440 BAYREUTH
Germany

Don Ihde, Dr.
The Philosophy Department
Harriman Hall 213
State University of New York
STONY BROOK, NY 11794
USA

Ann Johnson, Dr.
Department of Philosophy
University of South Carolina
COLUMBIA, SC 29208
USA

Tarja Knuuttila, Dr.
Department of History
University of Helsinki
P.O. Box 9
FI-00014 HELSINKI
Finland

Günter Küppers, Dr.
Institute for Science and
Technology Studies
Bielefeld University
P.O. Box 100131
D-33501 BIELEFELD
Germany
(home: Prinzenstrasse 3
D-33602 BIELEFELD, Germany)

Holger Lange, Priv.-Doz. Dr.
Norwegian Forest and
Landscape Institute
Høgskoleveien 8
N-1430 ÅS
Norway

Johannes Lenhard, Dr.
Institute for Science and
Technology Research, and
Department of Philosophy
Bielefeld University
P.O. Box 100131
D-33501 BIELEFELD
Germany

Erika Mattila, Research Officer
London School of Economics and
Political Science
Houghton Street
LONDON WC2A 2AE
UK

Martina Merz, Dr.
Observatoire Science, Politique et
Société
Université de Lausanne
rue de Bassenges 4
CH-1024 ECUBLENS
Switzerland

Arthur C. Petersen, Dr.
Netherlands Environmental
Assessment Agency (MNP)
P.O. Box 303
NL-3720 AH BILTHOVEN
The Netherlands

Terry Shinn, Professor Dr.
GEMAS
Maison des Sciences de l'Homme
54 boulevard Raspail
F-5006 PARIS
France

Eric Winsberg, PhD
Assistant Professor
Department of Philosophy
University of South Florida
4202 East Fowler Ave, FAO 226
TAMPA, FL 33620
USA

BIOGRAPHICAL NOTES

Marcel Boumans is Associate Professor of Methodology and History of Economics at the University of Amsterdam. His domain of research is marked by 3 M's: Model, Measurement, and Mathematics. He is author of *How Economists Model the World into Numbers* (Routledge, 2005), and is currently editing: *Measurement in Economics: a Handbook* (Elsevier, forthcoming). His current research explores a methodology of models that function as instruments in social science.

Michael Hauhs was educated in forestry and soil science at the University of Göttingen. His dissertation (1985) was on water and ion budgets of a small forested headwater catchment in the Harz mountains. He worked for two years as a post-doc at the Norwegian Institute of Water Research (NIVA) in Oslo. Habilitation in Soil Science in Göttingen (1991). Since 1992 he holds the chair of Ecological Modeling at the Bayreuth Institute for Terrestrial Ecosystem Research (BITÖK, University of Bayreuth). His fields of specialisation are ecosystem theory, forest growth models, hydrology, and artificial life.

Don Ihde is Distinguished Professor of Philosophy and Director of the Technoscience Research Group at Stony Brook University, New York, USA. He is the author of thirteen single-author books and editor of others. His *Technics and Praxis* (1979) is often cited as the first English language book on philosophy of technology; *Technology and the Lifeworld* (1990) and *Instrumental Realism* (1991) are also well known. Recent books include *Chasing Technoscience*, edited with Evan Selinger (2003), *Bodies in Technology* (2001), and *Expanding Hermeneutics: Visualism in Science* (1998).

Ann Johnson is an assistant professor at the Unviersity of South Carolina, where she has a joint appointment in history and philosophy. She studies the history and philosophy of engineering, particularly the ways new knowledge is produced by engineers. She is the author of *Designing ABS: Engineering Culture and the Production of Knowledge*, which is to be published in 2007 by Duke University Press. Her current research focuses on the role of simulations in engineering practice in the period since 1960.

Tarja Knuuttila is currently Research Fellow in the Department of Philosophy at the University of Helsinki, Finland. She holds master's degrees both in economics (Helsinki School of Economics) and in philosophy (Department of Social and Moral Philosophy, University of Helsinki), and she defended her PhD thesis in 2005 in the Department of Philosophy, University of Helsinki. She has worked several years as a researcher in a multidisciplinary research group doing science and technology studies in the Center for Activity Theory and Developmental Work Research, University of Helsinki. She has also given courses in philosophy, semiotics, and science and technology studies at the University of Helsinki. Her main research interests include scientific representation and modeling with a special emphasis on the various roles of technological artifacts in science.

Günter Küppers is retired since Mai 2004. He was educated in physics and worked on pattern formation in fluids at the Max Planck Institute for Plasma Physics.

Since 1974 he worked in the field of Science and Technology Studies at Bielefeld University. He wrote many books and articles on self-organisation, chaos, complexity, innovation networks, and computer simulations. He is teaching at the Universities of Bielefeld and Vienna.

Holger Lange was educated in physics and did his dissertation in theoretical physics on stability analysis of quantum field theoretic models 1992, University of Dortmund. 1992-2003 researcher at the Bayreuth Institute for Terrestrial Ecosystem Research (BITÖK), University of Bayreuth. Habilitation 1999 in Ecological Modeling, University of Bayreuth. Since 2003 senior researcher at the Norwegian Forest Research Institute (Skogforsk), Ås, Norway. His fields of specialisation are: ecological modeling, hydrology, systems theory, nonlinear dynamics and time series analysis.

Johannes Lenhard studied philosophy and mathematics at the universities of Heidelberg and Frankfurt. In 1998, he finished his dissertation in mathematics. Currently he is affiliated with the Institute for Science and Technology Studies (IWT) and the department of philosophy at Bielefeld University. His fields of interest cover the philosophy of science and science studies, in particular the subject of applied mathematics. During the last years a main focus of research is on epistemological and methodological issues of computer simulation. Recent publications include *Models and Statistical Inference: The Controversy between Fisher and Neyman-Pearson* (British Journal for the Philosophy of Science, 57, 2006) and *Surprised by a Nanowire: Simulation, Control, and Understanding* (Philosophy of Science, PSA 2004).

Martina Merz (Dr.) is senior researcher at the Observatory on Science, Policy and Society, University of Lausanne where she is responsible for the research area Social Studies of Science and Technology. She is also senior scientist at the Technology and Society Laboratory, EMPA, St. Gallen, and teaches sociology at the University of Lucerne.

Arthur Petersen (1970) is senior policy analyst and director of the Methodology and Modelling Programme at the Netherlands Environmental Assessment Agency (Milieu- en Natuurplanbureau. MNP). He leads the agency's efforts in methodology for sustainability assessment and methodology for uncertainty assessment and communication. He holds master's degrees in theoretical physics and philosophy of science from the Vrije Universiteit Amsterdam and a PhD in atmospheric science from Utrecht University. This year he expects his second PhD in philosophy of science (Vrije Universiteit). He is co-editor of *Sharing the Planet* (2003), *Not Afraid of Uncertainty* (2003, in Dutch) and *Remember Your Humanity* (2005). His book *Simulating Nature* will be published in 2006.

Erika Mattila is currently working as a research officer at London School of Economics in a Leverhulme Trust and ESRC funded project *Nature of Evidence: How Well Do 'Facts' Travel* at Economic History Department. Before that, Mattila worked in a project on *Multidisciplinary Modelling and Transformation of Scientific Work*, funded by the Helsinki Institute of Science and Technology Studies, at the Center for Activity Theory and Developmental Work Research,

University of Helsinki. Her research examines the construction and use of probabilistic modelling in the study of infectious diseases. The key findings enrich our understanding of the challenges faced in organising and conducting multidisciplinary research.

Terry Shinn is *Directeur de Recherche at the Centre National de Recherche Scientifique* (*CNRS*) and based at the *Paris Maison des Sciences de l'Homme*. His present work focuses on the contributions of a special form of instrumentation, generic instrumentation dubbed *research technologies*, to the paradoxical stabilization and transversality of the boundaries that simultaneously enshrine and suffocate scientific disciplines and other forms of social activity. Instrumentation is identified as a key mechanism of boundary conservation, boundary crossing, and cognitive commensuration, thereby preserving divisions of labor and also promoting a form of trans-organizational and trans-functional lingua franca and pragmatic universality. In this capacity, instrumentation is foundational both to scientific disciplines and to interdisciplinarity. Terry Shinn's recent publications include *Instrumentation Between Science, State and Industry* (2001) and *Controverses sur la Science: Pour une Sociologie Transversaliste de l'Activité Scientifique* (2005).

Eric Winsberg recieved his PhD in History and Philosophy of Science at Indiana University in 1999. He has been in the Philosophy Department at the University of South Florida since 2001 where he pursues research in philosophy of science, especially philosophy of physics. His primary research interests have been in the role of models and simulations in the sciences, and in the origins of time asymmetries and the foundations of statistical mechanics.

Name Index

Abraham, F.F. 140, 142–151
Adelman, F.L. 111–114, 122
Adelman, I. 111–114, 122
Allison, C. 202
Amico, V. 190, 191, 201
Anttila, A. 44, 55
Arakawa, A. 93–95, 99, 104
Arbab, F. 59, 61, 76
Attie, P. 76

Bachlechner, M.E. 151
Bailey, D. 38
Barlas, Y. 120–122
Batterman, R. 150, 151
Bednarek, A.R. 106
Bekkendahl, T. 38
Belitz, K. 105
Bernstein, N. 151
Bethune, D. 38
Beusen, A. 185
Beven, K. 57, 76
Bjerkenes, V. 59, 76
Black, E. 48, 54
Borchers, J. 104
Boumans, M. 126, 128, 135–137
Bourdieu, P. 5, 17, 21
Bousquet, F. 74, 76
Brooks, R. 65, 76
Broughton, J.Q. 141–145, 150, 151
Brown, J.H. 57, 77
Burns, A.F. 113, 114, 121, 122

Cambrosio, A. 127, 137
Campbell, T.J. 151
Carrier, M. 102, 104
Cartwright, N. 13, 21
Casti, J.L. 183, 184
Chelikowski, J. 151
Cho, K. 35, 39
Chomsky, N. 43, 45, 54
Christ, C.F. 111, 123
Chua, L.O. 13, 21
Collins, H.M. 48, 54
Community Climate System Model – CCSM 100, 101, 104
Cooley, T.F. 117, 118, 120, 123
Coopersmith, J. 31, 38
Coral Quintana, S. 185

Dai, H. 39
Daston, L. 137
De Kwaadsteniet, H. 173, 174, 183, 184
De Marchi, B. 185
De Vries, B. 185
Degan, L.A. 77
Dekker, C. 39
Demeritt, D. 104, 105
Dennett, D 79, 86
DeRose, S.J. 54
Devoret, M. 39
Dillon, A. 31, 38
Dolya, A.V. 194, 198, 202
Dowling, D. 155, 171
Drexler, K.E. 28, 38

Eady, E. 94, 105
Edwards, P.N. 104, 105
Evans, R. 48, 54
Franklin, A. 117, 123
French, S. 41, 54
Friedman, M. 114, 123
Frigg, R. 41, 54
Frisch, M. 148, 149, 151
Funtowicz, S.O. 174, 175, 178, 182, 184, 185

Galison, P.L. 18, 21, 32, 38, 42, 55, 80, 86, 104, 105, 171
Gardner, M. 12, 21
Geerligs, L. 39
Giere, R.N. 41, 55
Gieryn, T. 136, 137
Globus, A. 29, 38
Goddard, W. 38
Goldberger, A.S. 111–114, 121, 123
Goldin, D. 61, 76, 77
Griesemer, J. 127, 136, 138
Grim, P. 80, 81, 86
Guala, F. 137

Haag, D. 176, 184
Haas, P.M. 176, 178, 184
Haavelmo, T. 110, 111, 122, 123
Hacking, I. 41, 55
Hamada, N. 31, 38
Han, J. 38
Harmon, M.E. 77
Harremoës, P. 185
Hauhs, M. 57, 63, 68, 72, 76
Heben, M. 38
Heikkilä, J. 44, 55
Hertz, H. 107–109, 123
Heuberger, P. 185
Hoover, K.D. 115, 116, 123
Hoppe, R. 185
Hornberger, G.M. 66, 77
House, P. 202
Hughes, R.I.G. 41, 54, 55

Hughes, T.P. 28, 38
Huijs, S.W.F. 185
Humphreys, P. 7, 12, 21, 42, 50, 54, 55, 103–105

IEEE Computer 202
IEEE Transaction 191, 202
Ingold, T. 59, 72, 76
Intergovernmental Panel on Climate Change – IPCC 177–180, 182, 184, 185
International Vocabulary of Basic and General Terms in Metrology – IVM 107, 123

Jaffe, R. 38
Jakeman, A.J 66, 77
Janssen, P.H.M. 180, 184, 185
Järvinen, T. 54, 55
Jasanoff, S. 177, 178, 184, 185
Joerges, B. 19, 22, 193, 199, 202
Johnson, A. 8, 17, 25, 35, 38
Jones, K. 38

Kalev, D. 202
Kalia, R.K. 151
Karlsson, F. 44, 45, 48, 49, 54, 55
Kaupenjohann, M. 176, 184
Kaxiras, E. 151
Keating, P. 127, 137
Keller, E.F. 12, 20, 22, 104, 105, 155, 171
Kerr, R.A. 98, 105
Kiang, C. 38
Kimmins, H. 57, 77
King, R.G. 118, 121, 123
Klamer, A. 115, 123
Klein, L.R. 111, 112–114, 121, 123
Klein Thompson, J. 137
Kloprogge, P. 185
Knauft, F.-J. 76
Knorr-Cetina, K. 157, 159, 161, 163, 171
Knuuttila, T. 41, 50, 52, 55
Koch, J. 76
Koenig, A. 194, 202
Koskenniemi, K. 43, 48, 54, 55
Kratz, T.K. 67, 77
Krayer von Krauss, M.P. 185
Kremer, J.N. 77
Krück, C.C. 104, 105
Küppers, G. 22, 54, 89, 95, 102, 104, 105
Kydland, F.E. 108, 116–118, 123

Lange, H. 57, 76
Lansing, J.S. 58, 77
Latour, B. 157, 161, 169, 171
Lauenroth, W.K. 77
Laymon, R. 150, 151
Le Page, C. 74, 76
Leino, T. 137, 138
Lenhard, J. 3, 11, 14, 22, 54, 89, 95, 104, 105
Levit, C. 38
Lewis, J.M. 93, 104, 105
Lidorikis, E. 151
Lorenz, E. 104, 105
Lucas, R.E. 108, 114–117, 120, 123
Lücking, H. 104, 105
Lynch, M. 156, 171

Mann, T. 3, 22
Mar, G. 80, 86
Mattila, E. 125, 136–138
Maturana, H.R. 16, 22
Maxwell, J.C. 108, 123
McCann, K.S. 75, 77
McLeod, J. 190, 191, 201, 202
Meitner, M. 77
Menon, M. 35, 39
Merkle R. 27–30, 32, 34, 35, 38
Merz, M. 50, 55, 155, 156, 161, 163, 170, 171
Metropolis, N. 104, 105
Miller, C.A. 104, 105
Minelli, A. 70, 77
Mitchell, W.C. 113, 114, 121–123
Moreno, A. 77
Morgan, M.S. 12, 13, 22, 41, 50, 55, 111, 123, 136, 138, 155, 171, 172
Morrison, M. 13, 22, 41, 50, 55, 105, 138, 171, 172
Musgrave, C. 29, 38

Nakano, A. 140, 151
Nelson, E. 202
Neumann von, J. 91–93, 104, 105
Norton, S.D. 98, 105, 175, 185
Nuijten, D. 185

Ogata, S. 151
Orcutt, G.H. 120, 123
Oreskes, N. 98, 105
Oshiyama, A. 38

Parkinson, S. 106
Pedersen, E.K. 66, 77
Pereira, A.G. 185
Peretó, J. 77
Perry, J. 38
Peters, R.H. 57, 77
Petersen, A.C. 104, 105, 173, 177, 180, 182, 184, 185
Petroski, H. 38
Phillips, N. 93, 94, 105
Pickering, A. 41, 55, 84, 86
Pittroff, W. 66, 77

Plosser, C.I. 118, 121, 123
Potting, J. 185
Pradeepa Siva, C. 202
Prescott E.C. 108, 116–118, 120, 123
Pyka, A. 102, 105

Quintana, S.C. 185

Ragouet, P. 193, 199, 202
Ratner, M. 151
Ravetz, J.R. 174, 175, 178, 182, 184, 185
Rebelo, S.T. 118, 121, 123
Rheinberger, H.-J. 51, 55
Richtmyer, R.D. 105
Risbey, J.S. 185
Rob, R. 77
Rohrlich, F. 12, 22, 104, 105
Rosen, R. 58, 70, 75, 77
Roska, T. 13, 21
Rotmans, J. 185
Rubin, D.C. 64, 77
Rudd, R.E. 141, 150, 151
Ruiz-Mirazo, K. 76, 77
Rutten, J. 68, 77

Sawada, S. 38
Schellnhuber, H.J 59, 77
Schneider, S.H. 104, 105
Schütz, J.-P 57, 77
Schwartz, A.J. 114, 123
Schweber, S. 104, 106
Schwechheimer, H. 105
Science Citation Index (SCI®), Thomson ISI 5, 18, 22
Scott, P. 171, 172
Searle, J. 15, 22
Seely, B. 77
Shackley, S. 104, 106
Shieber, S. 15, 22
Shimojo, F. 151
Shinn, T. 3, 8, 19, 20, 22, 54, 187, 193, 199, 202
Shrader-Frechette, K. 105
Simon, H.A. 115, 116, 121–123
Sismondo, S. 155, 172
Smalley, R. 30, 39
Smolka, S.A. 76
Smuts, B. 77
Sonderegger, E. 76
Srivastava, D. 31, 32, 35–39
St. Denis, P. 86
Star Leigh, S. 138
Stillinger, F.H. 144, 146, 147, 151
Stokey, N.L. 120, 123
Stroustrup, B. 193–195, 197–199, 201–203
Suppe, F. 98, 105, 175, 185

Tans, S.J. 31, 39
Tapanainen, P. 46, 54, 55
Teller P. 41, 55
The Oxford English Dictionary, fourth edition, 1989 3, 22
Thess, A. 39
Thissen, W.A.H. 114, 123
Tom, S. 77
Tuinstra, W. 185
Turi, D. 76, 77
Turing, A.M. 13, 14, 16, 22, 108, 114, 115, 117, 121, 123

Ulam, S.M. 91, 92, 104–106
Ulanowicz, R.E. 58, 77

Van Asselt, M.B.A. 177, 183, 185
Van Daalen, C.E. 123
Van den Bogaard, A. 137, 138
Van der Sluijs, J.P. 180, 183–185
Van der Zwaan, B.C.C. 184, 185
Van Egmond, N.D. 173, 185
Van Fraassen, B.C. 123
Van Vuuren, D. 185
Vashishta, P. 151
Venner, B. 194, 195, 203
Verbraeck, A. 114, 123
Voss, D. 29, 39
Voutilainen, A 44, 45, 47–50, 52, 54, 55
Voyiadjis, G.Z. 151

Wächter, M. 106
Walker, W.E. 184, 185
Weber, T.A. 144, 146, 147, 151
Wegner, P. 61, 77
Weinberg, A.M. 175, 185
Welham, C. 77
Wenzel, V. 59, 77
West, G.B. 57, 77
White, K.P. 120, 124
Winsberg, E. 8, 13, 14, 22, 42, 54, 55, 99, 104, 106, 139, 150, 151, 171, 172
Woodward, J. 107, 124
Woolgar, S. 157, 171
Wynne, B. 106, 177, 178, 184, 185

Young, P. 106

Zeigler, B.P. 15, 22

Sociology of the Sciences

1. E. Mendelsohn, P. Weingart and R. Whitley (eds.): *The Social Production of Scientific Knowledge.* 1977 ISBN Hb 90-277-0775-8; Pb 90-277-0776-6
2. W. Krohn, E.T. Layton, Jr. and P. Weingart (eds.): *The Dynamics of Science and Technology.* Social Values, Technical Norms and Scientific Criteria in the Development of Knowledge. 1978 ISBN Hb 90-277-0880-0; Pb 90-277-0881-9
3. H. Nowotny and H. Rose (eds.): *Counter-Movements in the Sciences.* The Sociology of the Alternatives to Big Science. 1979 ISBN Hb 90-277-0971-8; Pb 90-277-0972-6
4. K.D. Knorr, R. Krohn and R. Whitley (eds.): *The Social Process of Scientific Investigation.* 1980 (1981) ISBN Hb 90-277-1174-7; Pb 90-277-1175-5
5. E. Mendelsohn and Y. Elkana (eds.): *Sciences and Cultures.* Anthropological and Historical Studies of the Sciences. 1981 ISBN Hb 90-277-1234-4; Pb 90-277-1235-2
6. N. Elias, H. Martins and R. Whitley (eds.): *Scientific Establishments and Hierarchies.* 1982 ISBN Hb 90-277-1322-7; Pb 90-277-1323-5
7. L. Graham, W. Lepenies and P. Weingart (eds.): *Functions and Uses of Disciplinary Histories.* 1983 ISBN Hb 90-277-1520-3; Pb 90-277-1521-1
8. E. Mendelsohn and H. Nowotny (eds.): *Nineteen Eighty Four: Science between Utopia and Dystopia.* 1984 ISBN Hb 90-277-1719-2; Pb 90-277-1721-4
9. T. Shinn and R. Whitley (eds.): *Expository Science.* Forms and Functions of Popularisation. 1985 ISBN Hb 90-277-1831-8; Pb 90-277-1832-6
10. G. Böhme and N. Stehr (eds.): *The Knowledge Society.* The Growing Impact of Scientific Knowledge on Social Relations. 1986 ISBN Hb 90-277-2305-2; Pb 90-277-2306-0
11. S. Blume, J. Bunders, L. Leydesdorff and R. Whitley (eds.): *The Social Direction of the Public Sciences.* Causes and Consequences of Co-operation between Scientists and Non-scientific Groups. 1987 ISBN Hb 90-277-2381-8; Pb 90-277-2382-6
12. E. Mendelsohn, M.R. Smith and P. Weingart (eds.): *Science, Technology and the Military.* 2 vols. 1988 ISBN Vol, 12/1 90-277-2780-5; Vol. 12/2 90-277-2783-X
13. S. Fuller, M. de Mey, T. Shinn and S. Woolgar (eds.): *The Cognitive Turn.* Sociological and Psychological Perspectives on Science. 1989 ISBN 0-7923-0306-7
14. W. Krohn, G. Küppers and H. Nowotny (eds.): *Selforganization.* Portrait of a Scientific Revolution. 1990 ISBN 0-7923-0830-1
15. P. Wagner, B. Wittrock and R. Whitley (eds.): *Discourses on Society.* The Shaping on the Social Science Disciplines. 1991 ISBN 0-7923-1001-2
16. E. Crawford, T. Shinn and S. Sörlin (eds.): *Denationalizing Science.* The Contexts of International Scientific Practice. 1992 (1993) ISBN 0-7923-1855-2
17. Y. Ezrahi, E. Mendelsohn and H. Segal (eds.): *Technology, Pessimism, and Postmodernism.* 1993 (1994) ISBN 0-7923-2630-X
18. S. Maasen, E. Mendelsohn and P. Weingart (eds.): *Biology as Society? Society as Biology: Metaphors.* 1994 (1995) ISBN 0-7923-3174-5
19. T. Shinn, J. Spaapen and V. Krishna (eds.): *Science and Technology in a Developing World.* 1995 (1997) ISBN 0-7923-4419-7

Sociology of the Sciences

20. J. Heilbron, L. Magnusson and B. Wittrock (eds.): *The Rise of the Social Sciences and the Formation of Modernity.* Conceptual Change in Context, 17501850. 1996 (1998) ISBN 0-7923-4589-4
21. M. Fortun and E. Mendelsohn (eds.): *The Practices of Human Genetics.* 1998 ISBN 0-7923-5333-1
22. B. Joerges and T. Shinn (eds.): *Instrumentation: Between Science, State and Industry.* 2000 ISBN 0-7923-6736-7
23. B. Joerges and H. Nowotny (eds.): *Social Studies of Science and Technology: Looking Back Ahead.* 2003 HB ISBN 1-4020-1481-3; Pb 1-4020-1482-1
24. S. Maassen and P. Weingart (eds.): *Democratization of Expertise. Exploring Novel* Forms of Scientific Advice in Political Decision- Making. 2005 HB ISBN 1-4020.3753-8; Pb 1-4020-3754-6
25. J. Lenhard, G. Küppers, T. Shinn (eds.): *Simulation. Pragmatic Construction of Reality.* 2006 ISBN 1-4020-5374-6